NO GRID SURVIVAL PROJECTS BIBLE

Transform Your Life with Proven DIY Strategies for Secure Living, Sustainable Food and Energy Independence – Your Blueprint to Thrive in Any Crisis or Economic Downturn

Alex Parker

Download the 5 bonus videos here by framing the QR code.

Copyright ©2024. All rights reserved.

NO GRID SURVIVAL PROJECTS BIBLE

All rights reserved

Table of Contents

Foreword 1
Introduction 2
PART I FOUNDATIONS OF NO-GRID LIVING 3
Chapter 1 Understanding No-Grid Survival 4
    The Concept and Philosophy 4
    Historical Evolution and Modern Relevance 5
    Myths vs. Reality: Debunking Common Misconceptions 6
Chapter 2 Setting the Mindset 8
    Emotional Preparedness 8
    Building Resilience and Adaptability 9
    Developing a Community Mindset 10
PART II ESSENTIAL SKILLS AND PREPARATION 12
Chapter 3 Basic Survival Skills 13
    Fire-Making Techniques and Safety 13
        Traditional Fire-Making Techniques: 13
        Fire Safety Measures 15
    Shelter-Building Fundamentals 15
        Lean-to Shelter 15
        Debris Hut Shelter 16
        A-Frame Shelter 17
        DIY Shelter-Building Tools 17
    Basic Navigation and Wilderness Orientation 19
        Map Reading and Compass Navigation 19
        Natural Navigation 20
        DIY Navigation Tools 20
        Wilderness Orientation Projects 21
        Wilderness Orientation Safety Measures 21
Chapter 4 Advanced Survival Techniques 22
    Long-Term Food and Water Sourcing 22
        Food Sourcing Techniques: 22
        DIY Food Sourcing Projects: 23
        Water Sourcing Techniques: 24
        DIY Water Sourcing Projects: 24
        Food and Water Sourcing Safety Measures: 25
    Adaptation to Various Climates and Terrain 26
        Adaptation Techniques 26
        DIY Projects for Climate and Terrain Adaptation: 27

Self-Defense and Safety in Remote Locations 28
   Situational Awareness 28
   Self-Defense Techniques 28
   Safety Measures for Remote Locations: 29
   DIY Projects for Personal Security: 29

PART III FOOD SECURITY AND AGRICULTURE 31

Chapter 5 Establishing Your Food Source 32

Planning Your Vegetable Garden 32
Principles of Horticulture and Crop Rotation 35
   Principles of Horticulture 35
   Crop Rotation 37
Managing Small-Scale Agriculture 38

Chapter 6 Livestock and Animal Husbandry 40

Choosing Livestock: Considerations and Care 40
   Step-by-Step Projects 41
Sustainable Animal Farming Practices 43
Processing and Utilizing Animal Products 44
   Meat Processing 44
   Milk Processing 45
   Egg Processing 45
   Wool Processing 45
   Utilizing Byproducts 46

Chapter 7 Foraging and Wild Edibles 47

Identifying Edible Plants and Mushrooms 47
   Edible Plants 47
   Edible Mushrooms 48
Ethical Foraging Practices 49
Incorporating Wild Foods into Your Diet 50

PART IV WATER MANAGEMENT 52

Chapter 8 Water Harvesting and Purification 53

Techniques for Collecting and Storing Water 53
   Rain Barrels 53
   Cisterns 53
   Catchment Ponds 54
DIY Filtration and Purification Methods 55
   Boiling 55
   Cloth Filtration 55
   Sand and Gravel Filtration 55
   Charcoal Filtration 56
   Solar Disinfection (SODIS) 56
   DIY Ceramic Filter 56

DIY Berkey Water Filter..56
Building and Maintaining Water Systems 57

Chapter 9 Sustainable Water Use 59
Conserving Water in Daily Activities 59
Implementing Greywater and Rainwater Systems 60
Greywater System..60
Rainwater Harvesting System...60
Managing Water in Various Climates 61

PART V SHELTER AND HOMEBUILDING 64

Chapter 10 Designing Your Off-Grid Home 65
Choosing the Right Location and Materials...65
Building Techniques for Durability and Efficiency 66
Customizing Your Living Space 67

Chapter 11 Energy-Efficient Home Solutions 69
Insulation and Passive Solar Design 69
Insulation...69
Passive Solar Design...69
Renewable Energy Options for Home Use 70
Maintaining and Upgrading Your Home 71

PART VI POWER INDEPENDENCE 73

Chapter 12 Solar Power Solutions..74
Basics of Solar Energy and Panel Installation 74
Battery Storage and Power Management 75
Types of Battery Storage Systems: 75
Power Management 76
Maximizing Efficiency in Solar Systems 76

Chapter 13 Alternative Energy Sources 78
Wind, Hydro, and Biomass Energy...78
Wind Energy..78
Hydroelectric Energy 78
Biomass Energy..79
DIY Energy Projects and Innovations...79
Balancing Multiple Energy Sources 80

PART VII COMMUNICATION AND COMMUNITY 82

Chapter 14 Off-Grid Communication 83
Establishing Reliable Communication Systems..83
Emergency Communication Protocols...84
Modern Communication Technologies...85

Chapter 15 Building a Supportive Community 88
    Creating Alliances and Sharing Resources 88
    Organizing Community Events and Workshops 89
    Collaborative Projects and Community Development 90

PART VIII HEALTH AND WELL-BEING 93

Chapter 16 Health Care and Emergency Preparedness 94
    Building a Comprehensive First Aid Kit 94
    Handling Medical Emergencies Off-Grid 95
        Cuts and Wounds 95
        Burns 95
        Fractures and Sprains 96
        Heat Exhaustion and Heatstroke 96
        Allergic Reactions 96
        Choking 96
        Heart Attack 96
        Seizures 96
    Mental Health Considerations in Isolation 97

Chapter 17 Mental Health and Social Aspects 99
    Coping with Isolation and Stress 99
    Building a Balanced Lifestyle 100
    Community Support for Mental Well-being 101

PART IX ADVANCED PROJECTS AND INNOVATIONS 103

Chapter 18 Specialized DIY Projects 104
    Building a Greenhouse and Aquaponics System 104
        Building a Greenhouse 104
        Building an Aquaponics System 105
    Advanced Water and Energy Systems 105
        Advanced Water System 105
        Advanced Energy System 106
    Smart Technologies in Off-Grid Living 107

Chapter 19 Leveraging Technology 108
    Using Drones and Sensors for Efficiency 108
    Automation in Off-Grid Systems 109
    Staying Updated with Emerging Technologies 110

PART X TRANSITION AND ADAPTATION 112

Chapter 20 Making the Transition 113
    Planning and Executing Your Move 113
    Overcoming Initial Challenges 114
    Financial Planning for Off-Grid Transition 115

Chapter 21 Future-Proofing Your Lifestyle 117
- Keeping Up with Innovations 117
- Preparing for Long-Term Sustainability 118
- Adapting to Environmental and Climate Changes 119

Conclusion 121
- Reflecting on the No-Grid Journey 121
- The Road Ahead 121

Appendices 123
- A: Essential Tools and Equipment 123
- B: Recommended Reading and Resources 124
- C: Checklists and Planning Guides 124

Index 127

Foreword

Welcome to a journey unlike any other—the exploration of no-grid living. As you embark on this adventure, allow me to share with you my personal journey into the world of off-grid living—a journey filled with challenges, discoveries, and profound moments of connection with nature and community.

For me, the decision to embrace no-grid living was born out of a deep longing for simplicity, authenticity, and sustainability. I found myself disillusioned with the fast-paced, consumer-driven lifestyle that seemed to disconnect me from the natural world and my own sense of purpose. Yearning for a more meaningful existence, I embarked on a quest to rediscover the fundamental principles of life—self-reliance, resilience, and harmony with nature.

My journey began with a leap of faith—a decision to leave behind the comforts of urban living and venture into the wilderness in search of a simpler way of life. Armed with little more than determination and a desire for change, I set out to build my own off-grid homestead—a place where I could live in harmony with the land, harnessing its resources responsibly and cultivating a deep sense of connection with the natural world.

What followed was a journey of discovery and transformation. I learned to harness the power of renewable energy, harvesting sunlight and wind to power my home and sustain my daily needs. I embraced the art of water conservation, capturing rainwater and managing resources with care. I cultivated a thriving garden, growing my own food and nurturing a connection with the cycles of the seasons.

But perhaps the most profound aspect of my journey into no-grid living was the sense of community that emerged along the way. I discovered that off-grid living is not a solitary endeavor but a collective experience—a journey shared with like-minded individuals who value sustainability, resilience, and connection. Together, we supported one another, shared knowledge and resources, and forged deep bonds of friendship and camaraderie.

As you embark on your own journey into no-grid living, I invite you to embrace the unknown, to welcome the challenges as opportunities for growth, and to open your heart to the transformative power of living in harmony with nature. May this guide serve as a beacon of inspiration and guidance as you navigate the path ahead, and may your journey be filled with discovery, connection, and fulfillment.

Here's to a life lived off-grid—a life of simplicity, sustainability, and profound connection with the natural world. Welcome to the adventure.

Introduction

In a world dominated by modern conveniences and interconnected technologies, the allure of off-grid living beckons as a path to simplicity, self-sufficiency, and harmony with nature. Embracing the off-grid lifestyle represents a departure from the conventional norms of urban living and a return to the roots of human existence, where reliance on the land and community fosters resilience and connection.

Off-grid living is more than just a lifestyle choice; it's a philosophy that challenges the status quo and redefines what it means to live a fulfilling and sustainable life. It's about reclaiming autonomy over our basic needs—energy, water, food—and forging a deeper connection with the natural world that sustains us.

Basically, embracing the off-grid lifestyle is a conscious decision to prioritize simplicity, sustainability, and self-reliance. It's about reducing our environmental footprint, conserving resources, and living in harmony with the Earth's natural rhythms. It's about disconnecting from the grid—both metaphorically and literally—and reconnecting with ourselves, our communities, and the planet.

In this book, we will explore several facets of off-grid living, from setting up renewable energy systems and harvesting rainwater to growing our own food and building resilient communities. We'll delve into the practicalities of transitioning to off-grid living, offering insights, tips, and resources to help you navigate this transformative journey.

Whether you're drawn to the tranquility of rural homesteading, the adventure of remote wilderness living, or the camaraderie of intentional communities, the off-grid lifestyle offers endless possibilities for personal growth, fulfillment, and connection.

Let's get started!

PART I
FOUNDATIONS OF NO-GRID LIVING

Chapter 1
Understanding No-Grid Survival

The term "no-grid survival," which is also often referred to as "off-grid living," describes a way of life in which individuals or communities strive to become self-sufficient and independent from public utilities such as electricity, water, and sewage systems. The generation of power from renewable energy sources, including solar panels, wind turbines, or hydroelectric generators, is frequently a component of this lifestyle choice. It may also involve collecting and purifying water from natural sources like rainwater, rivers, or wells, as well as managing waste through composting or other environmentally friendly methods.

The motivations for embracing a no-grid survival lifestyle can vary widely. Some people may choose this lifestyle for environmental reasons, seeking to reduce their carbon footprint and live in harmony with nature. Others may be drawn to it for economic reasons, aiming to save money on utility bills or escape the pressures of modern consumer culture. Additionally, some individuals may be motivated by a desire for greater self-reliance and resilience in the face of potential emergencies or disasters that could disrupt traditional infrastructure.

Living off-grid requires careful planning and preparation to ensure that basic needs such as food, water, shelter, and energy are met sustainably. This may involve learning traditional skills such as gardening, food preservation, and home repair, as well as adopting modern technologies and techniques for renewable energy and resource management.

While no-grid survival offers many benefits, including increased autonomy and a closer connection to nature, it also presents unique challenges and limitations. Off-grid living can require significant upfront investment in infrastructure and ongoing maintenance to ensure reliability and efficiency. Additionally, it may involve sacrifices in terms of comfort and convenience compared to life in a traditional urban or suburban setting.

The Concept and Philosophy

Embracing a lifestyle of no-grid survival entails a holistic approach centered on self-reliance and independence from conventional infrastructure. Rooted in a set of guiding principles and beliefs, this concept fosters resilience, sustainability, and a profound connection to nature. Here's a breakdown of its core elements:

- **Self-Sufficiency**: Basically, no-grid survival emphasizes the ability to meet one's basic needs without relying on external systems or resources. This includes generating energy, sourcing water, producing food, and managing waste independently.
- **Resilience**: No-grid survivalists aim to build resilience in the face of potential disruptions to traditional infrastructure, such as power outages, water shortages, or other emergencies. By expanding their sources of energy, water resources, and techniques of food production, they are better able to endure difficulties that they did not anticipate.

- **Sustainability**: Many practitioners of no-grid survival prioritize sustainability and environmental stewardship. This involves minimizing their ecological footprint by using renewable energy sources, conserving resources, and minimizing waste.
- **Connection to Nature**: Living off-grid often involves a deeper connection to the natural world. By relying on natural resources and living in harmony with the environment, individuals can develop a greater appreciation for the Earth's ecosystems and their role within them.
- **Simple Living**: No-grid survival promotes a simpler way of life, free from the excesses of consumer culture. By prioritizing essential needs over material possessions and embracing frugality, individuals can lead more fulfilling and meaningful lives.
- **Community and Cooperation**: While no-grid survival often emphasizes self-sufficiency, it also recognizes the importance of community and cooperation. Many off-grid communities rely on shared resources, knowledge, and labor to thrive collectively.
- **Personal Freedom**: For some, no-grid survival represents a quest for personal freedom and autonomy. By breaking away from mainstream society and its norms, individuals can chart their own course and live according to their own values and beliefs.
- **Preparedness**: No-grid survivalists often prioritize preparedness for potential disasters or emergencies. This may involve stockpiling supplies, developing emergency plans, and acquiring skills that are valuable in times of crisis.

Historical Evolution and Modern Relevance

The concept of living off-grid has historical roots dating back to early human civilizations when communities were largely self-sufficient and relied on natural resources for survival. However, the modern off-grid movement has evolved significantly, influenced by various historical, social, and technological factors:

1. **Back-to-the-Land Movement**: In the late 1960s and 1970s, the back-to-the-land movement emerged as a countercultural response to urbanization and consumerism. Many individuals and families sought to escape the city and live self-sufficiently on rural homesteads, often adopting off-grid practices.
2. **Renewable Energy Development**: Advances in renewable energy technologies, such as solar panels and wind turbines, have made off-grid living more feasible and sustainable. As the cost of these technologies has decreased and their efficiency has improved, more people have been able to generate their own power independent of the grid.
3. **Environmental Concerns**: Growing concerns about environmental degradation and climate change have led many people to reconsider their dependence on fossil fuels and conventional utilities. Off-grid living offers a way to decrease carbon emissions and minimize ecological impact by using renewable energy sources and practicing sustainable living.
4. **Resilience and Preparedness**: Natural disasters, economic instability, and other crises have highlighted the vulnerabilities of centralized infrastructure systems. Off-grid living provides a

means of increasing resilience and preparedness by diversifying energy sources, water supplies, and food production methods.

5. **Technological Innovation**: Advances in communication technology, such as satellite internet and mobile networks, have made it easier for off-grid individuals and communities to stay connected to the broader world while still maintaining independence from traditional utilities.

6. **Social and Cultural Shifts**: Changing social norms and values, including a growing interest in simplicity, minimalism, and self-reliance, have contributed to the resurgence of interest in off-grid living. Additionally, the rise of remote work and digital nomadism has enabled more people to embrace off-grid lifestyles without sacrificing income or connectivity.

In the modern context, off-grid living remains relevant for various reasons:

- **Environmental Sustainability**: Off-grid living offers a way to reduce environmental impact and live more sustainably by relying on renewable energy sources and minimizing resource consumption.
- **Resilience and Self-Sufficiency**: Off-grid communities are better equipped to withstand disruptions to centralized infrastructure, such as power outages or water shortages, by generating their own resources and managing them locally.
- **Personal Freedom and Autonomy**: Off-grid living provides individuals with the opportunity to live according to their values and beliefs, free from the constraints of mainstream society and consumer culture.
- **Community Building**: Off-grid communities often prioritize cooperation, mutual support, and shared resources, fostering strong social connections and a sense of belonging.

Myths vs. Reality: Debunking Common Misconceptions

Debunking common misconceptions about off-grid living can help provide a clearer understanding of the realities and challenges of this lifestyle. Here are some myths and their corresponding realities:

Off-grid living is primitive and uncomfortable.

While off-grid living may involve simpler amenities and a reduced reliance on modern conveniences, it doesn't necessarily mean sacrificing comfort. Many off-grid homes are equipped with modern appliances powered by renewable energy sources, and innovative design techniques can maximize comfort and efficiency.

Off-grid living is only for extremists or survivalists.

Off-grid living appeals to a diverse range of individuals, including environmentalists, homesteaders, retirees seeking tranquility, and families looking for a healthier lifestyle. It's not solely the domain of extremists but rather a lifestyle choice that aligns with various values and priorities.

Off-grid living is illegal or not allowed in most places.

Off-grid living is legal in many areas, although specific regulations regarding building codes, zoning laws, and environmental regulations may vary. In some cases, obtaining permits and meeting certain requirements may be necessary, but off-grid living is not inherently illegal.

Off-grid living is prohibitively expensive and only accessible to the wealthy.

While there can be initial costs associated with setting up off-grid infrastructure, such as solar panels or water filtration systems, off-grid living can ultimately be more cost-effective in the long run, as it reduces or eliminates monthly utility bills. Moreover, there are various ways to achieve off-grid living at different budget levels, from DIY solutions to community-based initiatives.

Off-grid living means complete isolation and disconnection from society.

Off-grid living doesn't necessarily mean isolation; many off-grid communities are interconnected and supportive. Additionally, technological advancements such as internet access and mobile communication enable off-grid individuals to stay connected to the broader world while still enjoying the benefits of a self-sufficient lifestyle.

Off-grid living is impractical and unsustainable in the long term.

With careful planning, resource management, and technological innovation, off-grid living can be highly sustainable and resilient. Many off-grid communities practice permaculture, water conservation, and renewable energy generation, minimizing their ecological footprint and ensuring long-term viability.

Off-grid living requires extreme self-sufficiency and expertise in survival skills.

While some level of self-sufficiency is necessary for off-grid living, it doesn't require extreme expertise or survival skills. Many off-grid individuals and communities learn as they go, acquiring knowledge and skills over time through experience, research, and community support.

Chapter 2
Setting the Mindset

Embarking on the path to self-reliance begins with the right mindset—a mindset of preparedness, adaptability, and community. Setting the mindset involves embracing a shift in perspective towards self-sufficiency, sustainability, and resilience. It requires recognizing the interconnectedness of our actions with the environment and prioritizing simplicity and resourcefulness over excess and consumption. This mindset entails a deep appreciation for nature and a commitment to living in harmony with it, valuing the intrinsic worth of natural resources and seeking to minimize our ecological footprint. It also involves cultivating a sense of personal responsibility for one's well-being and the well-being of the community, fostering a spirit of cooperation and mutual support.

Emotional Preparedness

Emotional preparedness is a vital aspect of readiness for any significant life change, including transitioning to off-grid living. While the focus often falls on the practical aspects of such a lifestyle change, it's crucial not to overlook the emotional journey that accompanies it. Emotional preparedness involves understanding and addressing the range of feelings, challenges, and adjustments that may arise as one embarks on this new way of life.

First and foremost, it's essential to acknowledge that transitioning to off-grid living can bring about a mix of emotions, including excitement, anticipation, anxiety, and even fear. Leaving behind the familiarity and comfort of mainstream society and stepping into the unknown can be daunting. However, embracing these emotions as natural responses to change is the first step towards emotional preparedness.

One of the most significant emotional challenges in transitioning to off-grid living is the sense of isolation or loneliness that may arise, especially if one is moving to a remote location. Living off-grid often means being physically distant from neighbors, friends, and family members, which can lead to feelings of disconnection. It's essential to realise these feelings and actively work to build and maintain connections with others, whether through regular communication, participation in community events, or joining local groups and organizations.

Another emotional aspect to consider is the sense of loss or mourning for the conveniences and comforts of modern life that may be left behind. This could include access to unlimited electricity, running water, and easy access to goods and services. It's normal to grieve these losses and to feel a sense of nostalgia for the familiar, but it's also an opportunity to reevaluate priorities and cultivate gratitude for the simplicity and beauty of off-grid living.

Moreover, transitioning to off-grid living may require letting go of certain expectations or preconceived notions about what life should look like. This could involve redefining success, happiness, and fulfillment in terms that align with the values of sustainability, self-sufficiency, and community. In order to successfully navigate this change, it is vital that one approaches it with a willingness to learn and a readiness to adjust to new methods of thinking and living.

Emotional preparedness also involves developing resilience and coping strategies to navigate the inevitable challenges and setbacks that may arise along the way. Living off-grid requires a certain degree of self-reliance and problem-solving skills, as well as the ability to weather storms, power outages, and other unforeseen circumstances. The cultivation of a positive outlook, the practice of self-care, and the solicitation of assistance from others when it is required are all components of building resilience.

Additionally, it's crucial to manage expectations and maintain a realistic perspective on the challenges and limitations of off-grid living. While it offers many rewards, including a closer connection to nature, greater autonomy, and a simpler way of life, it also comes with its share of hardships and sacrifices. It's essential to approach off-grid living with a balanced outlook, acknowledging both the joys and the challenges that come with it.

Finally, emotional preparedness involves embracing the sense of adventure and possibility that comes with embarking on a new journey. Transitioning to off-grid living is an opportunity for personal growth, exploration, and self-discovery. It's a chance to reconnect with the natural world, cultivate meaningful relationships, and live with intention and purpose. By approaching this transition with an open heart and a sense of curiosity, one can embrace the full richness and potential of off-grid living.

Building Resilience and Adaptability

Resilience and adaptability are two sides of the same coin, essential qualities for anyone pursuing a self-reliant lifestyle. Resilience is the ability to withstand adversity, to bounce back from challenges stronger than before. Adaptability, on the other hand, is the capacity to adjust and thrive in ever-changing circumstances. Together, they form a formidable duo that can see you through the toughest of times. Here's how to cultivate these qualities:

1. **Develop a Positive Mindset**: Develop a mindset that is optimistic and resilient by concentrating on finding solutions instead of obsessing about the challenges that you are facing. Practice gratitude and mindfulness to maintain a positive outlook even in challenging situations.
2. **Build a Supportive Community**: It is important to surround yourself with people who share your values and who are able to offer you support, encouragement, and practical aid when you require it. Strengthening social connections and fostering a sense of belonging can enhance resilience and provide a safety net during difficult times.
3. **Learn from Challenges**: View setbacks and failures as opportunities for growth and learning. Consider your past experiences to recognize the lessons they hold and to formulate strategies for surmounting similar challenges in the future.
4. **Stay Flexible and Adaptable**: Embrace change and uncertainty as inherent aspects of off-grid living. Remain open to new ideas, perspectives, and ways of doing things, and be willing to adjust your plans and expectations as circumstances evolve.
5. **Develop Practical Skills**: Acquire a diverse range of practical skills that are relevant to off-grid living, such as gardening, food preservation, renewable energy systems, and emergency preparedness. Building competence in these areas enhances self-sufficiency and increases confidence in your ability to handle challenges.
6. **Prioritize Self-Care**: It is important to prioritize self-care methods like regular exercise, proper sleep, good food, and stress management strategies in order to ensure that your mental, physical,

and emotional health are treated with the utmost importance. Maintaining a strong foundation of self-care enhances resilience and fosters adaptability.

7. **Plan for Contingencies**: To reduce possible hazards and ensure that you are well-prepared, it is important to plan for potential issues and build contingency plans. Consider factors such as extreme weather events, equipment failures, and supply shortages, and have backup strategies in place to address these scenarios.

8. **Embrace Innovation and Creativity**: Be open to experimenting with new ideas, technologies, and approaches to problem-solving. Coming up with new ideas and being creative are really important for solving tricky problems and adjusting well when things change.

9. **Celebrate Progress and Achievements**: Acknowledge and celebrate your successes, no matter how small, to maintain motivation and momentum. Recognizing your accomplishments boosts confidence and resilience, providing a sense of accomplishment and fueling continued progress.

10. **Stay Connected to Nature**: Cultivate a deep connection to the natural world and draw strength from its beauty, resilience, and abundance. Spending time outdoors, immersing yourself in nature, and observing its cycles can provide perspective and inspiration during challenging times.

Developing a Community Mindset

While self-reliance often conjures images of solitude and independence, the reality is that no one truly thrives alone. Developing a community mindset is not just about building connections for mutual benefit; it's about recognizing the inherent strength that comes from collective support. Here's how to cultivate a community mindset in the context of off-grid living:

1. **Promote Open Communication**: Establish channels for open and transparent communication within the community, allowing members to express their thoughts, concerns, and ideas freely. Regular meetings, online forums, and social gatherings provide opportunities for dialogue and collaboration.

2. **Encourage Collaboration**: Foster a culture of collaboration and cooperation by encouraging members to work together on common goals and projects. Shared tasks such as gardening, infrastructure maintenance, and emergency preparedness build solidarity and strengthen community bonds.

3. **Share Resources**: Embrace the principle of resource-sharing within the community, pooling resources such as tools, equipment, and supplies for the benefit of all members. By sharing resources, individuals can reduce costs, increase efficiency, and enhance resilience in the face of challenges.

4. **Offer Mutual Support**: Be willing to offer and receive support from fellow community members during times of need. Whether it's lending a helping hand with a construction project, providing emotional support during difficult times, or sharing surplus food from the garden, acts of kindness and generosity strengthen community bonds and foster a sense of belonging.

5. **Celebrate Diversity**: Embrace the diversity of skills, backgrounds, and perspectives within the community, recognizing that each member brings unique strengths and experiences to the table. Celebrate cultural differences, encourage inclusivity, and create an environment where everyone feels valued and respected.

6. **Foster a Spirit of Empowerment**: Empower community members to take initiative, make decisions, and contribute to the collective well-being of the group. Encourage leadership development, delegate responsibilities, and provide opportunities for individuals to pursue their passions and interests within the community.

7. **Practice Conflict Resolution**: Develop constructive strategies for resolving conflicts and addressing disagreements within the community. Encourage open dialogue, active listening, and compromise, seeking mutually beneficial solutions that preserve relationships and foster understanding.

8. **Engage in Shared Learning**: Create opportunities for shared learning and skill-building within the community, such as workshops, classes, and hands-on demonstrations. By learning together, community members can expand their knowledge, develop new skills, and strengthen bonds through shared experiences.

9. **Cultivate Resilience**: Foster a culture of resilience within the community, emphasizing the importance of preparedness, adaptability, and self-reliance. Encourage members to work together to identify and mitigate risks, develop contingency plans, and support one another during times of crisis.

10. **Nurture a Sense of Belonging**: Create a welcoming and inclusive environment where every member feels a sense of belonging and connection. Celebrate milestones, acknowledge contributions, and foster traditions and rituals that reinforce the shared identity and collective spirit of the community.

PART II

ESSENTIAL SKILLS AND PREPARATION

Chapter 3
Basic Survival Skills

In the wilderness, where modern conveniences are stripped away, basic survival skills become paramount. This chapter serves as your essential primer on fundamental techniques that can mean the contrast between life and death in the wild. From mastering fire-making to constructing shelters and navigating through unknown terrain, these skills form the backbone of wilderness survival knowledge.

Fire-Making Techniques and Safety

Fire is more than just a source of warmth—it's a lifeline in the wilderness. Knowing how to start a fire from scratch is a skill that can mean the difference between comfort and discomfort, or even life and death.

Traditional Fire-Making Techniques:

Friction Fire Methods

1. **Bow Drill**

The bow drill is a classic friction fire-making technique that involves using a spindle, a hearth board, a bow, and a socket. Here's how to make and use a bow drill:

Materials Needed:

- Hardwood spindle (e.g., oak, maple)
- Softwood hearth board (e.g., cedar, pine)
- Flexible bow (e.g., green sapling, paracord)
- Socket (e.g., stone, bone, hardwood)

Procedure:

1. Carve a divot into the hearth board and create a notch leading from the divot to the edge of the board.
2. Attach one end of the bowstring to the bow and wrap it around the spindle.
3. Place the spindle into the divot on the hearth board and secure the socket on top of the spindle.
4. Use the bow to rotate the spindle rapidly, creating friction between it and the hearth board.
5. Continue rotating the bow until the friction generates enough heat to create an ember in the notch.
6. Carefully transfer the ember to a tinder bundle and blow gently to ignite the tinder.
7. Once the tinder ignites, carefully transfer it to a prepared fire lay to build your fire.

2. **Hand Drill**

The hand drill method is similar to the bow drill but involves rotating the spindle between your hands instead of using a bow. Here's how to make and use a hand drill:

Materials Needed:

- Hardwood spindle
- Softwood hearth board
- Handhold (e.g., a small piece of wood or stone)

Procedure:

1. Carve a divot into the hearth board and create a notch leading from the divot to the edge of the board.
2. Hold the spindle between your hands and place it into the divot on the hearth board.
3. Apply downward pressure and rotate the spindle rapidly by rolling it between your palms.
4. Continue rotating the spindle until the friction generates enough heat to create an ember in the notch.
5. Transfer the ember to a tinder bundle and blow gently to ignite the tinder.
6. Once the tinder ignites, transfer it to a prepared fire lay to build your fire.

DIY Fire-Starting Tools:

1. **Fire Piston**

A fire piston is a compact and efficient fire-starting tool that uses the principle of rapid compression to ignite tinder. Here's how to make a simple fire piston:

Materials Needed:

- Metal or plastic tube (e.g., brass tubing, PVC pipe)
- Solid plunger rod (e.g., wooden dowel, metal rod)
- O-ring or rubber gasket
- Char cloth or other tinder material

Procedure:

1. Cut a piece of tubing to the desired length for your fire piston.
2. Drill a small hole in one end of the tubing to accommodate the plunger rod.
3. Insert the plunger rod into the tubing, ensuring a snug fit.
4. Install the O-ring or rubber gasket around the plunger rod near the open end of the tubing to create an airtight seal.
5. To use the fire piston, place a small piece of char cloth or other tinder material in the open end of the tubing.
6. Quickly and forcefully push the plunger rod into the tubing, compressing the air inside.
7. The rapid compression generates enough heat to ignite the tinder, creating an ember that can be transferred to a tinder bundle to make a fire.

2. **Fire Steel and Flint**

A fire steel and flint set is a traditional fire-starting tool that produces sparks when struck together. Here's how to make a DIY fire steel and flint set:

Materials Needed:

- High-carbon steel rod (e.g., hacksaw blade, file)

- Flint or quartz rock
- Striker (e.g., knife blade, sharp-edged metal)

Procedure:

1. Cut a piece of high-carbon steel rod to the desired length for your fire steel.
2. Grind or file one end of the steel rod to create a sharp edge.
3. Select a piece of flint or quartz rock with a sharp edge for striking.
4. Hold the flint firmly in one hand and strike the edge of the flint with the sharp edge of the fire steel.
5. The impact produces sparks, which can be directed onto a tinder bundle to ignite it and start a fire.

Fire Safety Measures

- **Clear the Area:** Before starting a fire, clear the surrounding area of any flammable materials, such as dry leaves, grass, or branches, to prevent the fire from spreading uncontrollably.
- **Build a Fire Lay:** Construct a proper fire lay using small, dry tinder at the base, followed by progressively larger kindling and fuelwood arranged in a teepee, log cabin, or other suitable configuration.
- **Keep Water Nearby:** Have a ready supply of water, such as a bucket or hose, nearby to extinguish the fire in case it gets out of control.
- **Never Leave a Fire Unattended:** Always supervise the fire and never leave it unattended, especially in windy conditions or when children or pets are present.
- **Completely Extinguish the Fire:** When you're done with the fire, ensure that it is completely extinguished by dousing it with water and stirring the ashes 'til they are cool to the touch.

Shelter-Building Fundamentals

A well-constructed shelter is a sanctuary against the elements, offering protection from harsh weather conditions and preserving body heat. In survival situations, knowing how to build shelters using natural materials is an invaluable skill.

Lean-to Shelter

The lean-to shelter is one of the simplest and most versatile shelter designs, consisting of a sloped roof supported by a ridgepole and anchored to the ground at one end. Here's how to build a lean-to shelter:

Materials Needed:

- Ridgepole (long, sturdy branch or pole)
- Support poles or trees
- Branches, sticks, or foliage for roofing
- Rope or cordage (optional)

Procedure:

1. Find a suitable location for your lean-to shelter, preferably with natural features such as trees or rock formations to serve as support for the ridgepole.
2. Prop one end of the ridgepole against a sturdy tree or anchor it securely to the ground using support poles.
3. Extend the ridgepole diagonally to the ground, creating a sloped roof angle that provides protection from rain and wind.
4. Secure the ridgepole in place by tying it to nearby trees or anchoring it with additional support poles if necessary.
5. Collect branches, sticks, or foliage to lay across the ridgepole, creating a roof covering that deflects rain and provides insulation.
6. Thatch the roof covering tightly together to minimize gaps and improve weatherproofing.
7. Add additional insulation and waterproofing layers if available, such as leaves, pine boughs, or a tarp, to enhance comfort and protection.
8. Clear the ground inside the shelter of any debris or sharp objects, and create a raised bed or sleeping platform using branches or foliage for added insulation and comfort.

Debris Hut Shelter

The debris hut shelter is a more advanced shelter design that provides better insulation and protection from the elements. It consists of a framework of support poles covered with layers of foliage, leaves, or other natural materials. Here's how to build a debris hut shelter:

Materials Needed:

- Long, sturdy support poles
- Branches, sticks, or saplings for framework
- Leaves, foliage, or grass for insulation
- Rope or cordage (optional)

Procedure:

1. Find a suitable location for your debris hut shelter, preferably with ample natural materials for construction and insulation.
2. Prop two long support poles upright in the ground to serve as the main framework for the shelter.
3. Lean additional branches, sticks, or saplings against the support poles at an angle, creating a ribbed framework for the shelter walls.
4. Fill in the gaps between the framework with smaller branches, foliage, or grass to create a dense, insulated layer.
5. Continue adding layers of insulation until the shelter walls are thick and sturdy enough to provide adequate protection from the elements.
6. Create a low entrance at one end of the shelter, leaving enough space for easy entry and exit while minimizing heat loss.
7. Place a thick layer of insulation to the floor of the shelter using leaves, grass, or other soft materials for comfort and warmth.

8. Optional: Secure the framework and insulation layers in place using rope or cordage, tying knots to anchor the materials securely.

A-Frame Shelter

The A-frame shelter is a simple yet effective shelter design that provides ample protection from rain and wind. It consists of two support poles or trees with a ridgepole laid across them, forming an A-shaped frame. Here's how to build an A-frame shelter:

Materials Needed:

- Two sturdy support poles or trees
- Ridgepole (long, sturdy branch or pole)
- Branches, sticks, or foliage for roofing
- Rope or cordage (optional)

Procedure:

1. Find a suitable location for your A-frame shelter, ensuring that there are two sturdy support poles or trees spaced apart at an appropriate distance.
2. Prop the support poles upright in the ground or anchor them securely to trees using rope or cordage.
3. Lay the ridgepole horizontally across the top of the support poles, forming an A-shaped frame with equal angles on each side.
4. Secure the ridgepole in place by tying it to the support poles or trees using rope or cordage, ensuring a stable and secure connection.
5. Collect branches, sticks, or foliage to lay across the ridgepole, creating a roof covering that deflects rain and provides insulation.
6. Thatch the roof covering tightly together to minimize gaps and improve weatherproofing.
7. Add additional insulation and waterproofing layers if available, such as leaves, pine boughs, or a tarp, to enhance comfort and protection.
8. Clear the ground inside the shelter of any debris or sharp objects, and create a raised bed or sleeping platform using branches or foliage for added insulation and comfort.

DIY Shelter-Building Tools

Bushcraft Knife

A bushcraft knife is a versatile and essential tool for shelter-building, as well as for various other outdoor tasks such as carving, cutting, and processing wood. Here's how to make a simple DIY bushcraft knife:

Materials Needed:

- High-carbon steel blade blank
- Wooden handle material (e.g., hardwood, antler)
- Epoxy or wood glue
- Sandpaper or file
- Cord or leather for handle wrapping (optional)

Procedure:

1. Select a high-quality blade blank made of high-carbon steel, with a tang that extends into the handle for strength and durability.
2. Shape the wooden handle material to fit the contours of your hand comfortably, ensuring a secure and ergonomic grip.
3. Use epoxy or wood glue to attach the blade blank to the wooden handle, ensuring a tight and secure bond.
4. Allow the epoxy or glue to dry completely according to the manufacturer's instructions.
5. Sand or file the handle to smooth out any rough edges and create a comfortable grip.
6. Optional: Wrap the handle with cord or leather for added grip and comfort, securing the wrapping with epoxy or glue.
7. Sharpen the blade to a razor-sharp edge using a sharpening stone or file, ensuring that it is suitable for cutting and carving tasks.

Camp Axe

A camp axe is another essential tool for shelter-building, chopping wood, and processing materials in an outdoor environment. Here's how to make a simple DIY camp axe:

Materials Needed:

- Axe head
- Wooden handle material (e.g., hardwood, hickory)
- Epoxy or wood glue
- Wedges or metal pins for securing the handle
- Sandpaper or file
- Leather or cord for handle wrapping (optional)

Procedure:

1. Select a high-quality axe head made of durable steel, with a sharp cutting edge and a sturdy construction.
2. Choose a suitable wooden handle material, such as hardwood or hickory, with a straight grain and minimal defects.
3. Insert the axe head into the top of the wooden handle, ensuring a snug and secure fit.
4. Use epoxy or wood glue to bond the axe head to the wooden handle, filling any gaps or voids to create a strong and durable connection.
5. Insert wedges or metal pins into the top of the handle to secure the axe head in place, ensuring a tight and stable fit.
6. Allow the epoxy or glue to dry completely according to the manufacturer's instructions.
7. Sand or file the handle to smooth out any rough edges and create a comfortable grip.
8. Optional: Wrap the handle with leather or cord for added grip and comfort, securing the wrapping with epoxy or glue.
9. Sharpen the cutting edge of the axe head to a razor-sharp edge using a sharpening stone or file, ensuring that it is suitable for chopping and cutting tasks.

Shelter-Building Safety Measures:
- **Site Selection:** Choose a shelter-building site that is flat, well-drained, and free from hazards such as falling branches, dead trees, or rocky terrain.
- **Materials Collection:** Gather shelter-building materials responsibly, avoiding damage to living trees, plants, or wildlife habitats. Use dead or fallen branches and foliage whenever possible.
- **Hydration and Rest:** Stay hydrated and well-rested during the shelter-building process to maintain energy levels and prevent fatigue or injury.
- **Tool Safety:** Handle sharp tools such as knives and axes with care, following proper safety precautions to prevent accidents or injuries.
- **Weather Awareness:** Monitor weather conditions closely and be prepared to seek shelter or adjust your plans accordingly in the event of adverse weather.
- **Leave No Trace:** Reduce your environmental impact by minimizing disturbance to the surroundings and ensuring your shelter-building site is as pristine or improved upon as when you arrived.

Basic Navigation and Wilderness Orientation

Being able to navigate the wilderness without relying on GPS or maps is a fundamental survival skill. Whether you're lost, disoriented, or simply exploring off the beaten path, understanding basic navigation techniques can guide you to safety.

Map Reading and Compass Navigation

Map Reading

Map reading involves understanding and interpreting topographic maps to navigate accurately in the wilderness. Here's how to read a topographic map:
1. **Understanding Map Symbols:** Familiarize yourself with common map symbols and features such as contour lines, landmarks, trails, and water bodies.
2. **Orienting the Map:** Align the map with the surrounding terrain using a compass or by matching known landmarks with their corresponding features on the map.
3. **Identifying Terrain Features:** Use contour lines to visualize the elevation and terrain features such as hills, valleys, ridges, and depressions.
4. **Determining Distance and Direction:** Use the map's scale to estimate distances between landmarks and use the compass to determine directions of travel.

Compass Navigation

Compass navigation involves using a compass in conjunction with a map to navigate accurately and efficiently. Below are steps on how to effectively use a compass for navigation:
1. **Setting the Compass:** Orient the compass by aligning the direction of travel arrow with the desired direction on the map or terrain.
2. **Taking Bearings:** Use the compass to take bearings from known landmarks or features and transfer them to the map to determine your location or direction of travel.

3. **Following Bearings:** Follow the compass needle to maintain the desired direction of travel, periodically checking the map to ensure you're on course.
4. **Adjusting for Declination:** Adjust for magnetic declination by either setting the declination on the compass or mentally adding or subtracting the declination angle from your bearings.

Natural Navigation

Natural navigation involves using natural features such as the sun, stars, moon, and landscape to determine direction and navigate without relying on compasses or maps. Here are some natural navigation techniques:

1. **Solar Navigation:** Use the sun's position in the sky to determine direction. In the Northern Hemisphere, the sun ascends from the east and descends in the west, mirroring its pattern in the Southern Hemisphere where it also rises from the east and sets in the west.
2. **Stellar Navigation:** Use prominent stars, such as Polaris (the North Star), and recognizable constellations to determine direction. Polaris can be found in the northern sky and serves as a reliable reference point for navigation in the Northern Hemisphere.
3. **Moon Navigation:** Use the moon's phases and position in the sky to determine direction. In the Northern Hemisphere, the illuminated side of the moon faces south, while in the Southern Hemisphere, it faces north.

DIY Navigation Tools

DIY Compass

A simple DIY compass can be made using a piece of cork, a magnetized needle, and a container of water. Here's how to make a DIY compass:

1. **Magnetize the Needle:** Rub a needle against a magnet repeatedly in one direction to magnetize it.
2. **Float the Needle:** Place the magnetized needle on a small piece of cork and float it in a container of water.
3. **Wait for Alignment:** Hold your breath and wait for the needle to align itself with the magnetic field of the Earth, which will indicate the direction of north to south.
4. **Mark the Direction:** Mark the north-south direction on the container to use as a reference for navigation.

Sun Dial

A sun dial is a simple DIY navigation tool that uses the sun's shadow to indicate the time and approximate direction. Here's how to make a sun dial:

1. **Choose a Dial Plate:** Select a flat surface, such as a piece of wood or stone, to serve as the dial plate.
2. **Insert a Gnomon:** Attach a vertical stick or rod (gnomon) to the center of the dial plate, ensuring that it is perpendicular to the plate's surface.
3. **Mark Hour Lines:** Use a compass or known landmarks to determine the cardinal directions (north, south, east, west) and mark hour lines on the dial plate accordingly.

4. **Place the Sun Dial:** Position the sun dial in a sunny location with the gnomon pointing north (in the Northern Hemisphere) or south (in the Southern Hemisphere).
5. **Read the Time:** Use the shadow cast by the gnomon to indicate the time based on the hour lines marked on the dial plate.

Wilderness Orientation Projects

- **Map and Compass Exercise:** Practice map reading and compass navigation skills by planning and executing a hiking or orienteering course in a local park or wilderness area. Use a topographic map and compass to navigate to predetermined checkpoints or landmarks, adjusting for terrain features and obstacles along the way.
- **Night Navigation Challenge:** Test your navigation skills at night by embarking on a night hike or orienteering challenge using only natural navigation techniques such as star and moon navigation. Practice identifying prominent stars and constellations to determine direction and navigate through the darkness with confidence.

Wilderness Orientation Safety Measures

- **Plan Ahead:** Plan your route carefully and familiarize yourself with the terrain, landmarks, and potential hazards before setting out.
- **Inform Others:** In the event that something unexpected occurs, you should keep someone informed of your intended route, the time you anticipate returning, and the emergency contact information.
- **Carry Essential Gear:** Bring essential navigation tools such as maps, compasses, GPS devices, and communication devices (e.g., cell phone, two-way radio) for safety and navigation.
- **Stay Oriented:** Continuously monitor your surroundings and refer to your navigation tools to maintain awareness of your location and direction of travel.
- **Stay Calm:** Remain calm and composed if you become disoriented or lost, and take steps to retrace your steps, find a reference point, or seek assistance if needed.

Chapter 4
Advanced Survival Techniques

In the journey of self-reliance and resilience, mastering basic survival skills is crucial. However, as you delve deeper into the world of survivalism, you'll encounter situations that demand more advanced techniques. This chapter explores advanced survival techniques that go beyond the basics, equipping you with the knowledge and skills needed to thrive in challenging environments and prolonged emergencies.

Long-Term Food and Water Sourcing

Long-term food and water sourcing are crucial aspects of off-grid living, self-sufficiency, and wilderness survival. Establishing sustainable methods for procuring food and water ensures ongoing access to essential resources and enhances resilience in diverse environments.

Food Sourcing Techniques:

Gardening and Permaculture

Gardening and permaculture techniques involve cultivating food crops, herbs, and perennial plants in a sustainable and regenerative manner. Here's a step-by-step guide to help you embark on your permaculture gardening journey:

1. **Site Selection:** Choose a sunny location with well-drained soil and access to water for your garden.
2. **Design the Garden:** Design your garden layout using permaculture principles such as polyculture planting, companion planting, and vertical gardening to maximize productivity and diversity.
3. **Prepare the Soil:** Improve soil fertility and structure by adding compost, mulch, and organic amendments to enrich the soil and support plant growth.
4. **Select Suitable Crops:** It is important to select a wide range of varieties of vegetables, fruits, herbs, and perennial plants that are suitable for the climate and growth circumstances of your area.
5. **Plant and Maintain:** Plant seeds or seedlings according to the recommended spacing and planting dates, and maintain the garden by watering, weeding, and pest control as needed.
6. **Harvest and Preserve:** Harvest ripe produce regularly and preserve excess harvest through methods such as canning, drying, fermenting, and freezing for long-term storage.

Foraging and Wildcrafting

Foraging and wildcrafting involve harvesting wild edible plants, mushrooms, and herbs from natural environments. Here's how to forage for wild edibles responsibly:

1. **Learn Plant Identification:** Educate yourself on plant identification, habitat, and seasonality to recognize edible and medicinal plants safely and accurately.
2. **Research Local Flora:** Research local foraging laws, regulations, and ethical guidelines to ensure responsible harvesting and conservation of wild plants and ecosystems.

3. **Start Small:** Begin by foraging for easily identifiable and common wild edibles such as dandelions, wild berries, and edible mushrooms, and gradually expand your knowledge and skills over time.
4. **Harvest Ethically:** Harvest wild plants sustainably by taking only what you need, leaving ample resources for wildlife and future generations, and avoiding harvesting rare or endangered species.
5. **Prepare and Preserve:** Clean, prepare, and cook wild edibles properly to ensure safety and palatability, and preserve excess harvest through methods such as drying, pickling, or infusing for long-term storage.

DIY Food Sourcing Projects:

Raised Bed Garden

Constructing raised bed gardens offers a practical and compact solution for cultivating vegetables, herbs, and flowers within a managed space. Below are steps to create your own raised bed garden:

1. **Materials Needed:** Gather materials such as untreated lumber, cedar boards, or recycled pallets for the raised bed frame, and organic soil mix for filling the beds.
2. **Construct the Frame:** Assemble the raised bed frame to your desired dimensions using screws or nails, ensuring a sturdy and level structure.
3. **Fill with Soil:** Fill the raised bed frame with organic soil mix, compost, and amendments to create a fertile growing medium for plants.
4. **Planting:** Plant seeds or seedlings in the raised bed according to spacing and planting guidelines, and provide adequate water and sunlight for plant growth.
5. **Maintenance:** Maintain the raised bed garden by watering, weeding, and fertilizing as needed, and monitor plant health for pests and diseases.
6. **Harvest:** Harvest ripe produce regularly and replant or rotate crops to maximize yield and productivity over time.

Rainwater Harvesting System

A rainwater harvesting system collects then stores rainwater for use in gardening, irrigation, and household tasks. Here's how to build a simple rainwater harvesting system:

1. **Materials Needed:** Gather materials such as a rain barrel or storage tank, gutters, downspouts, and a filtration system.
2. **Install Gutters and Downspouts:** Attach gutters and downspouts to your roof to collect rainwater and channel it into the storage tank or rain barrel.
3. **Connect Filtration System:** Install a filtration system or mesh screen at the entry point of the downspout to remove debris and contaminants from the collected rainwater.
4. **Position Rain Barrel:** Position the rain barrel or storage tank in a suitable location near your garden or outdoor area, ensuring stability and accessibility for watering.
5. **Collect Rainwater:** Allow rainwater to collect in the storage tank or rain barrel during rainfall events, and close the lid to prevent evaporation and contamination.
6. **Use for Irrigation:** Use collected rainwater for watering plants, gardens, and landscape features, reducing reliance on municipal water sources and conserving water resources.

Water Sourcing Techniques:

Natural Springs and Wells

Natural springs and wells provide access to groundwater for drinking, irrigation, and livestock. Here's how to locate and develop a natural spring or well:

1. **Survey the Landscape:** Survey the landscape for signs of water sources such as lush vegetation, wetland areas, or depressions in the terrain where groundwater may collect.
2. **Test for Water Quality:** Test water quality using a water testing kit or by sending samples to a certified laboratory for analysis to ensure safety and potability.
3. **Construct Well or Spring Box:** Construct a well or spring box structure to protect the water source from contamination and facilitate access for drawing water.
4. **Install Pump or Hand Pump:** Install a pump or hand pump mechanism to extract water from the well or spring box for use in drinking, irrigation, and household tasks.
5. **Maintain and Monitor:** Regularly maintain and monitor the well or spring box to ensure proper function, water quality, and integrity of the structure.

Rainwater Collection and Storage

Rainwater collection and storage systems capture and store rainwater for numerous uses such as irrigation, drinking, and household tasks. Here's how to set up a rainwater collection and storage system:

1. **Design the System:** Design a rainwater collection and storage system based on your available space, water needs, and local regulations, taking into account factors such as roof area, rainfall intensity, and storage capacity.
2. **Install Collection Surfaces:** Install gutters, downspouts, and collection surfaces such as roofs or awnings to channel rainwater into storage tanks or barrels.
3. **Choose Storage Containers:** Choose suitable storage containers such as rain barrels, cisterns, or underground tanks to store collected rainwater, ensuring they are clean, watertight, and UV-resistant.
4. **Filter and Treat:** Install filtration and treatment systems to remove debris, sediment, and contaminants from collected rainwater, ensuring it is safe and potable for drinking and other uses.
5. **Maintain and Monitor:** Regularly maintain and monitor the rainwater collection and storage system to ensure proper function, water quality, and cleanliness, cleaning filters, and disinfecting storage containers as needed.

DIY Water Sourcing Projects:

Berkey Water Filter

A Berkey water filter is an effective and portable filtration system for purifying drinking water in off-grid and emergency situations. Here's how to make a DIY Berkey water filter:

1. **Materials Needed:** Gather materials such as two food-grade plastic buckets, Berkey water filter elements, a spigot, and a lid.
2. **Prepare the Buckets:** Drill holes in the bottom of one bucket and in the lid of the other bucket to accommodate the Berkey filter elements and spigot.

3. **Assemble the Filter:** Insert the Berkey filter elements into the holes drilled in the bottom of the first bucket, ensuring they fit snugly and securely.
4. **Stack the Buckets:** Place the bucket with the filter elements on top of the second bucket with the lid, aligning the holes to allow water to flow through the filters.
5. **Install the Spigot:** Install a spigot at the bottom of the second bucket to dispense purified water, ensuring it is watertight and secure.
6. **Fill and Filter:** Fill the top bucket with unfiltered water and allow it to pass through the Berkey filter elements by gravity, collecting purified water in the bottom bucket for drinking and cooking.

Solar Still

A solar still is an improvised water distillation device that extracts clean water from contaminated sources such as saltwater, brackish water, or contaminated freshwater. Here's how to build a solar still:

1. **Materials Needed:** Gather materials such as a clear plastic sheet, a digging tool, a collection container, and rocks or weights.
2. **Dig a Hole:** Dig a hole in the ground large enough to accommodate the collection container and with sloping sides to funnel condensation towards the center.
3. **Place the Collection Container:** Place the collection container in the center of the hole to collect condensed water vapor.
4. **Cover with Plastic Sheet:** Cover the hole with a clear plastic sheet, ensuring it is taut then sealed around the edges to avoid moisture from escaping.
5. **Secure the Edges:** Secure the edges of the plastic sheet with rocks or weights to hold it in place and create a sealed chamber.
6. **Wait for Condensation:** Wait for the heat of the sun to evaporate moisture from the ground and plants inside the still, condensing on the underside of the plastic sheet and dripping into the collection container.
7. **Collect Purified Water:** Collect purified water from the collection container periodically, replenishing the moisture in the still as needed to continue the distillation process.

Food and Water Sourcing Safety Measures:

- **Test and Treat:** Test food and water sources for safety and quality, and treat or purify them as needed to remove contaminants and pathogens.
- **Rotate and Refresh:** Rotate food and water supplies regularly to ensure freshness and prevent spoilage, and refresh stored water periodically to maintain quality.
- **Monitor and Maintain:** Regularly monitor and maintain food gardens, water sources, and storage systems to ensure proper function, hygiene, and sustainability.
- **Diversify and Supplement:** Diversify food sources and water collection methods to enhance resilience and adaptability in changing environmental conditions.
- **Practice and Prepare:** Continuously practice food and water sourcing techniques, and prepare for emergencies by stockpiling essential supplies and developing contingency plans.

Adaptation to Various Climates and Terrain

Surviving in the wild often means encountering diverse climates and challenging landscapes. Whether you find yourself in a dense forest, arid desert, snowy tundra, or rugged mountains, adapting to your surroundings is essential for your well-being.

Understanding Climates and Terrains

- **Climates:** Climates can be broadly categorized into tropical, temperate, arid, and polar climates, each with distinct weather patterns, temperature ranges, and precipitation levels. Understanding the characteristics of different climates is crucial for planning and preparing for survival in various environments.
- **Terrains:** Terrains encompass landscapes such as mountains, forests, deserts, and coastal regions, each presenting unique challenges and opportunities for survival. Terrain features such as elevation, vegetation, water sources, and wildlife impact shelter-building, navigation, and resource procurement strategies.

Adaptation Techniques

Shelter-Building

Adapting shelter-building techniques to different climates and terrains is essential for protection from the elements and maintaining comfort and safety. Here are adaptation techniques for various environments:

- **Tropical Climates:** In tropical climates with high temperatures and humidity, prioritize ventilation and protection from insects and rain. Build elevated shelters with thatched roofs and open sides to maximize airflow while providing shade and protection from rain.
- **Arid Climates:** In arid climates with hot temperatures and limited water sources, focus on sheltering from the sun and conserving water. Construct low, insulated shelters with thick walls to minimize heat transfer and use shade structures such as awnings or lean-tos to provide relief from the sun.
- **Temperate Climates:** In temperate climates with moderate temperatures and seasonal variations, build versatile shelters that provide insulation and protection from rain and wind. Use materials such as wood, stone, or clay for sturdy construction and incorporate features such as windows, doors, and chimneys for comfort and functionality.
- **Polar Climates:** In polar climates with freezing temperatures and harsh winds, prioritize insulation and wind resistance in shelter construction. Build low, compact shelters with thick walls and insulated roofs, and use materials such as snow, ice, or earth for added insulation and protection from the cold.

Clothing and Gear

Adapting clothing and gear to different climates and terrains is essential for staying comfortable and safe in outdoor environments. Here are adaptation techniques for clothing and gear:

- **Layering:** Layer clothing to regulate body temperature and adapt to changing weather conditions. Wear moisture-wicking base layers to keep skin dry, insulating mid-layers to trap heat, and waterproof outer layers to repel rain and wind.
- **Footwear:** Choose footwear suitable for the terrain, such as hiking boots with sturdy soles and ankle support for rugged terrain, lightweight trail shoes for dry and flat terrain, and insulated boots for cold and snowy conditions.
- **Shelter Systems:** Invest in versatile shelter systems such as tents, hammocks, or bivouacs that can be adapted to different climates and terrains. Choose lightweight and durable materials with features such as waterproof coatings, insect screens, and ventilation panels for comfort and protection.
- **Tools and Equipment:** Select tools and equipment suited to the tasks and challenges of specific environments. Pack essentials such as knives, multi-tools, fire-starting devices, and navigation instruments, along with specialized gear for activities such as climbing, skiing, or water sports.

DIY Projects for Climate and Terrain Adaptation:

Tropical Shelter

Build a tropical shelter using bamboo, palm fronds, and other natural materials found in tropical environments. Here's how to construct a simple lean-to shelter:

1. **Gather Materials:** Collect bamboo poles, palm fronds, and other flexible materials for shelter construction.
2. **Frame the Shelter:** Prop up several bamboo poles against a sturdy tree or anchor them in the ground to form the frame of the shelter.
3. **Cover with Palm Fronds:** Lay palm fronds or large leaves across the bamboo frame to create a roof covering, overlapping them to shed rainwater.
4. **Secure and Insulate:** Secure the palm fronds in place with vines or cordage, and add additional layers for insulation and waterproofing if necessary.
5. **Clear the Ground:** Clear the ground inside the shelter of any debris or vegetation, and create a raised platform using bamboo or branches for sleeping and storage.

Desert Clothing

Create desert-appropriate clothing using lightweight and breathable fabrics that provide protection from the sun and sand. Here's how to make a desert headscarf:

1. **Choose Fabric:** Select a lightweight and breathable fabric such as cotton or linen in a neutral or light color to reflect sunlight and minimize heat absorption.
2. **Cut Fabric:** Cut the fabric into a large square or rectangle, approximately 36 to 48 inches on each side, depending on your preference.
3. **Fold and Tie:** Fold the fabric diagonally to form a triangle, and tie the ends around your head, leaving the point of the triangle hanging down to cover your neck and face.
4. **Adjust and Secure:** Adjust the fit and coverage of the headscarf as needed, and secure it in place with knots or tucks to prevent it from slipping or unraveling in the wind.

Mountain Shelter

Build a mountain shelter using rocks, logs, and other sturdy materials found in alpine environments. Here's how to construct a basic A-frame shelter:

1. **Collect Materials:** Gather rocks, logs, and branches for shelter construction, focusing on sturdy and stable materials.
2. **Build Frame:** Construct a triangular frame using two large logs or branches as the sides and a shorter log or branch as the ridgepole, forming an A-shaped structure.
3. **Fill In:** Fill in the gaps between the frame with smaller branches, rocks, or sod to create walls that provide insulation and wind protection.
4. **Cover Roof:** Lay additional branches or sod across the ridgepole to form a roof covering, overlapping them to shed rain and snow.
5. **Insulate and Seal:** Add insulation and waterproofing layers such as leaves, moss, or tarpaulin to the roof and walls, and seal any gaps or openings with additional materials to keep out wind and precipitation.

Self-Defense and Safety in Remote Locations

Remote and isolated environments present unique challenges, including encounters with wildlife, potential threats from human intruders, and unforeseen dangers. Advanced survival techniques encompass strategies for protecting oneself and ensuring safety in these scenarios.

Situational Awareness

- **Understanding Threats:** Identify potential threats in remote locations, including wildlife encounters, natural hazards, and human adversaries. Understanding the risks inherent in different environments is crucial for developing effective self-defense and safety strategies.
- **Environmental Awareness:** Be aware of your surroundings at all times, paying attention to terrain features, weather conditions, and signs of wildlife activity. Maintaining situational awareness lets you to anticipate potential dangers and take proactive measures to stay safe.
- **Trusting Instincts:** Trust your instincts and intuition when assessing unfamiliar situations or encountering unfamiliar individuals. If something feels off or unsafe, listen to your gut instinct and take appropriate action to remove yourself from the situation.

Self-Defense Techniques

Basic Self-Defense Moves

Learn basic self-defense techniques to defend yourself from physical threats and attackers. Here are some fundamental self-defense moves:

1. **Striking:** Practice basic striking techniques such as punches, kicks, and elbow strikes to create distance and incapacitate attackers.
2. **Blocking:** Learn blocking techniques to deflect and neutralize incoming attacks, protecting vital areas of the body from harm.

3. **Joint Locks and Holds:** Study joint locks and holds to control and restrain attackers, applying pressure to sensitive joints to subdue them without causing permanent injury.
4. **Escape and Evasion:** Practice escape and evasion tactics to remove yourself from dangerous situations, including techniques for breaking free from grabs, holds, and chokeholds.

Improvised Weapons

Use improvised weapons found in the environment to defend yourself in emergency situations. Common improvised weapons include:

1. **Sticks and Branches:** Use sturdy sticks or branches as clubs or batons to fend off attackers and create distance.
2. **Rocks and Stones:** Throw rocks or stones at attackers to distract and disorient them, buying time to escape or seek help.
3. **Sharp Objects:** Use sharp objects such as knives, broken glass, or pointed sticks as defensive weapons to inflict injury on attackers if necessary.

Safety Measures for Remote Locations:

- **Travel Preparation:** Plan and prepare for remote travel by researching the area, assessing potential risks, and packing essential supplies and equipment for self-defense and safety.
- **Communication Devices:** Carry communication devices such as satellite phones, two-way radios, or personal locator beacons to signal for help and stay connected in remote areas with limited cell coverage.
- **Emergency Signaling:** Learn and practice emergency signaling techniques such as using whistle blasts, signal mirrors, or brightly colored clothing to attract attention and summon assistance in remote locations.
- **Fire and Shelter Building:** For the purpose of providing warmth, shelter, and protection from the elements in the event of an emergency, acquire the skills necessary to build shelters and light fires. Constructing a fire and shelter can increase survival chances in remote environments.
- **Navigation and Orientation:** Develop proficiency in navigation and orientation techniques using maps, compasses, and natural landmarks to find your way in remote terrain and avoid getting lost.

DIY Projects for Personal Security:

Personal Alarm System

Create a DIY personal alarm system using simple electronic components to alert others in case of emergency. Here's how to make a basic personal alarm:

1. **Materials Needed:** Gather materials such as a small buzzer or piezoelectric buzzer, a push-button switch, a battery holder, and wires.
2. **Assemble the Circuit:** Connect the buzzer, switch, and battery holder in series to create a simple electrical circuit. Ensure proper connections and polarity.
3. **Enclose the Components:** Enclose the circuit components in a small plastic or metal enclosure to protect them from damage and weather.

4. **Activate the Alarm:** Carry the personal alarm with you and activate it by pressing the push-button switch in case of emergency, emitting a loud and audible alarm signal to attract attention.

Pepper Spray or Mace

Prepare homemade pepper spray or mace using natural ingredients for self-defense against human and animal threats. Here's a simple recipe for homemade pepper spray:

1. **Ingredients Needed:** Gather ingredients such as hot peppers (e.g., jalapenos, habaneros), rubbing alcohol or vodka, and a small spray bottle.
2. **Prepare the Peppers:** Finely chop or blend the hot peppers, removing seeds and membranes if desired to adjust the spiciness level.
3. **Mix with Alcohol:** Combine the chopped peppers with rubbing alcohol or vodka in a small container, allowing the peppers to infuse into the alcohol solution.
4. **Strain and Fill Bottle:** Strain the pepper-infused alcohol solution through a fine mesh sieve or cheesecloth to remove solid particles, and transfer the liquid into a small spray bottle for easy dispensing.
5. **Label and Store:** Label the homemade pepper spray or mace bottle clearly with its contents and usage instructions, and store it in a secure and easily accessible location for self-defense purposes.

Personal Defense Tool

Craft a simple personal defense tool using common materials found in nature or household items. Here's how to make a basic self-defense keychain:

1. **Materials Needed:** Gather materials such as a small metal rod or wooden dowel, a length of paracord or strong cord, and a key ring.
2. **Form Handle:** Wrap one end of the metal rod or wooden dowel with paracord to create a comfortable and secure handle grip, securing the wrapping with knots or glue.
3. **Attach Key Ring:** Tie or loop the key ring through the opposite end of the rod or dowel, creating a functional and discreet self-defense tool that can be carried on a keychain or lanyard.

Safety Measures and Practices

- **Avoid Confrontation:** Whenever possible, avoid confrontation and conflict by de-escalating tense situations, maintaining a calm demeanor, and seeking alternative solutions to disputes.
- **Stay Vigilant:** Remain vigilant and observant of your surroundings, looking out for potential threats and suspicious behavior in remote locations.
- **Trust Your Instincts:** Trust your instincts and intuition when assessing unfamiliar situations or encountering unfamiliar individuals, and take appropriate action to protect yourself if necessary.
- **Seek Assistance:** In case of emergency or imminent danger, seek assistance from authorities, nearby individuals, or rescue services as soon as possible.
- **Practice Regularly:** Regularly practice self-defense techniques, emergency procedures, and safety measures to maintain proficiency and readiness for remote environments.

PART III
FOOD SECURITY AND AGRICULTURE

Chapter 5
Establishing Your Food Source

When it comes to self-reliance and sustainability, few things are as fundamental as growing your own food. In this chapter, we delve deep into the art and science of establishing a reliable food source that sustains you through all seasons. By doing these things, you become less dependent on buying food from stores and more self-sufficient. This helps ensure that you always have enough to eat, especially during difficult times or emergencies.

Planning Your Vegetable Garden

A well-planned vegetable garden serves as the heart of self-sufficient living, providing fresh produce for sustenance and enhancing the resilience of your homestead. Here, we explore the essential steps to create a thriving garden that yields bountiful harvests year-round.

1. **Choose a Suitable Location**

Picking the best spot for your veggie garden is super important for it to thrive. Here are some things to think about when deciding where to put it:

- **Sunlight:** Select a site that receives a minimum of 6-8 hours of sunlight daily, as the majority of vegetables need abundant sunlight to flourish.
- **Soil Quality:** Assess the soil quality in potential garden sites, opting for well-drained, fertile soil with a pH level of 6.0-7.0 for optimal plant growth.
- **Accessibility:** Ensure easy access to your garden for weeding, watering, and harvesting, placing it near your home or water source if possible.
- **Space:** Determine the available space for your garden, considering factors such as the size of your property, existing landscaping, and future expansion possibilities.

2. **Design Your Garden Layout**

Designing a well-organized garden layout helps optimize space, improve efficiency, and facilitate maintenance tasks. Follow these steps to plan your garden layout:

- **Sketch Your Garden:** Use graph paper or a garden planning tool to sketch a layout of your garden, including dimensions, pathways, and bed locations.
- **Plan Bed Arrangement:** Arrange garden beds in a manner that maximizes sunlight exposure and airflow, avoiding shading from nearby structures or vegetation.
- **Consider Companion Planting:** Incorporate companion planting principles by grouping compatible plants together to enhance growth, deter pests, and improve overall garden health.
- **Include Pathways:** Design pathways between garden beds to provide access for maintenance tasks and prevent soil compaction.

- **Account for Watering:** Plan for efficient watering by positioning beds near water sources or installing irrigation systems as needed.

3. **Prepare the Soil**

Preparing the soil is essential for providing a healthy growing environment for your vegetables. Follow these steps to prepare your garden soil:

- **Clear the Area:** Remove any existing vegetation, rocks, or debris from the garden site to create a clean planting area.
- **Loosen the Soil:** Using garden fork or tiller to loosen the soil to a depth of 8-12 inches, breaking up compacted soil and improving drainage.
- **Amend with Organic Matter:** By incorporating organic matter into the soil, like compost, aged manure, or leaf mulch, you can strengthen the structure of the soil, boost the amount of moisture that is retained, and boost the fertility of the soil.
- **Test Soil pH:** Conduct a soil pH test using a soil testing kit to determine the acidity or alkalinity of the soil, adjusting pH levels as needed with amendments such as lime or sulfur.

4. **Select Vegetable Varieties**

Choosing the right vegetable varieties is essential for successful gardening, taking into account factors such as climate, growing season, and personal preferences. Consider the following when selecting vegetable varieties:

- **Climate Suitability:** Choose vegetable varieties adapted to your local climate and growing conditions, selecting heat-tolerant varieties for warm climates and cold-hardy varieties for cooler regions.
- **Growing Season:** Select vegetable varieties with appropriate maturity dates for your growing season, choosing early, mid-season, and late-season varieties for continuous harvests.
- **Space Requirements:** Consider the space requirements of each vegetable variety when planning your garden layout, spacing plants according to their mature size to prevent overcrowding.
- **Disease Resistance:** Choose disease-resistant vegetable varieties whenever possible to minimize the risk of pest and disease problems in your garden.

5. **Start Seeds or Transplants**

Once you've selected your vegetable varieties, it's time to start seeds or purchase transplants for planting in your garden. Follow these steps to start seeds or acquire transplants:

- **Seed Starting:** Seeds should be started indoors four to six weeks before the date that your region experiences its final frost. Seed trays or containers that are filled with seed-starting mix should be used.
- **Transplanting:** Purchase vegetable transplants from a reputable nursery or garden center, selecting healthy, vigorous plants with strong stems and well-developed roots.

- **Harden Off Seedlings:** Get your young plants used to being outside by letting them spend more time in the sun and wind each day. Do this for about a week to ten days before moving them into your garden.
- **Planting:** Plant seeds or transplants in your garden according to spacing and planting instructions provided on seed packets or plant labels, ensuring adequate spacing between plants for optimal growth.

6. **Implement Companion Planting**

Companion planting is a gardening technique that involves growing compatible plants together to enhance growth, repel pests, and improve overall garden health. Consider the following companion planting combinations for your vegetable garden:

- **Tomatoes, Basil, and Marigolds:** Plant basil and marigolds near tomatoes to deter pests such as aphids, whiteflies, and hornworms while enhancing tomato flavor.
- **Carrots, Onions, and Radishes:** Interplant carrots, onions, and radishes to maximize space and minimize pest problems, with onions deterring carrot flies and radishes repelling root maggots.
- **Beans and Corn:** Grow beans and corn together in a classic Native American planting combination known as the "Three Sisters," with beans fixing nitrogen in the soil for corn while corn provides support for beans to climb.
- **Lettuce and Herbs:** Plant lettuce and herbs such as parsley, dill, or chives together to attract beneficial insects and add flavor to salads while deterring pests.

7. **Implement Raised Beds and Container Gardening**

Raised beds and container gardening are versatile options for growing vegetables in limited space or poor soil conditions. Follow these steps to implement raised beds and container gardening in your vegetable garden:

- **Build Raised Beds:** Construct raised beds using lumber, bricks, or other materials, ensuring a depth of 6-12 inches for adequate root growth.
- **Fill with Soil:** Fill raised beds with a mixture of topsoil, compost, and organic matter to provide a fertile growing medium for vegetables.
- **Planting:** Plant vegetables directly in raised beds according to spacing and planting instructions, grouping compatible plants together for optimal growth.
- **Container Gardening:** Use containers such as pots, planters, or buckets to grow vegetables in small spaces or on patios and balconies, selecting containers with adequate drainage holes and suitable sizes for plant roots.

8. **Install Support Structures**

Many vegetables benefit from support structures such as trellises, stakes, or cages to encourage upright growth, improve airflow, and maximize space. Follow these steps to install support structures in your vegetable garden:

- **Identify Support Needs:** Determine which vegetables require support structures based on their growth habits and vine-like tendencies, such as tomatoes, cucumbers, and peas.
- **Choose Support Materials:** Select appropriate support materials such as bamboo stakes, wooden trellises, or wire cages, ensuring they are sturdy enough to withstand plant weight and wind.
- **Install Supports:** Place support structures in the garden bed or container before planting, securing them firmly in the soil or attaching them to nearby structures for stability.
- **Train and Tie Plants:** Train vines and tendrils to climb support structures as they grow, gently tying them with twine or plant ties to encourage upward growth and prevent sprawling.

9. **Maintain and Care for Your Garden**

Taking care of your vegetable garden regularly is really important to make sure it stays healthy and gives you lots of tasty veggies all season long. Follow these guidelines for maintaining your garden:

- **Watering:** Provide consistent and adequate watering to ensure plants receive enough moisture for healthy growth, watering deeply and infrequently to encourage deep root development.
- **Weeding:** Keep garden beds free of weeds by regularly removing unwanted vegetation that competes with plants for nutrients, water, and sunlight.
- **Mulching:** A layer of organic mulch, such as wood chips, straw, or shredded leaves, should be applied to garden beds in order to prevent the growth of weeds, maintain the soil's moisture content, and control the temp. of the soil.
- **Fertilizing:** Provide plants with necessary nutrients by applying organic fertilizers such as compost, aged manure, or organic plant food according to package instructions.
- **Pest and Disease Management:** Monitor plants regularly for signs of pests and diseases, implementing integrated pest management (IPM) strategies such as handpicking pests, using insecticidal soap, or applying organic pesticides as needed.
- **Harvesting:** Harvest vegetables at peak ripeness for best flavor and nutritional value, picking them regularly to encourage continuous production and prevent overripening or spoilage.

Principles of Horticulture and Crop Rotation

Horticulture, the art and science of cultivating plants, forms the backbone of successful gardening and agriculture. Understanding the principles of horticulture not only ensures a bountiful harvest but also promotes the long-term health of the soil and ecosystem. One of the fundamental practices in horticulture is crop rotation, a time-tested method that has been employed by farmers for centuries to maintain soil fertility and minimize pest and disease pressure.

Principles of Horticulture

1. **Soil Health**

Healthy soil is the foundation of successful horticulture, providing essential nutrients, water, and support for plant growth. Principles of soil health include:

- **Soil Structure:** Good soil structure promotes root penetration, water infiltration, and air circulation, facilitating plant growth and nutrient uptake.
- **Soil Fertility:** Fertile soil possesses sufficient quantities of vital nutrients like phosphorus, nitrogen, and potassium, essential for the growth and flourishing of plants.
- **Soil pH:** Soil pH affects nutrient availability and microbial activity, with most plants preferring a slightly acidic to neutral pH range of 6.0-7.0.
- **Organic Matter:** Incorporating organic matter such as manure, compost, or cover crops improves soil structure, fertility, and microbial activity, enhancing overall soil health.

2. **Plant Selection and Placement**

Choosing suitable plants and placing them appropriately in the garden are key principles of horticulture. Considerations include:

- **Climate and Microclimates:** Select plants adapted to your local climate and microclimates within your garden, taking into account factors such as sunlight, temperature, and rainfall.
- **Plant Diversity:** Include a variety of plant species, including vegetables, fruits, herbs, flowers, and native plants, to promote biodiversity, attract beneficial insects, and deter pests.
- **Companion Planting:** Plant compatible species together to enhance growth, repel pests, and improve overall garden health, following principles of companion planting.

3. **Water Management**

Effective water management is essential for maintaining plant health and optimizing water use efficiency. Principles include:

- **Watering Techniques:** Using techniques such as drip irrigation or soaker hoses, plants should be watered in a manner that is both deep and infrequent. This will promote the formation of deep roots and reduce the amount of water that is lost due to evaporation.
- **Mulching:** It is possible to maintain soil moisture, reduce weeds, and regulate soil temperature by applying organic mulch to garden beds. Some examples of organic mulch are straw, wood chips, and compost.
- **Rainwater Harvesting:** Collect and store rainwater for garden irrigation using rain barrels, cisterns, or swales, reducing reliance on municipal water sources and conserving freshwater resources.

4. **Pest and Disease Management**

Keeping pests and diseases away from plants is really important. It helps plants stay healthy and makes sure we don't lose too much of our crops. Principles of pest and disease management include:

- **Integrated Pest Management (IPM):** Implement a holistic approach to pest control that combines cultural, mechanical, biological, and chemical methods to minimize pest damage while minimizing environmental impacts.
- **Crop Rotation:** Rotate crops annually to disrupt pest and disease cycles, reduce soil-borne pathogens, and maintain soil fertility, following principles of crop rotation.

Crop Rotation

Crop rotation means planting different kinds of crops in a specific order on the same land every year. It's a way to keep the soil healthy and productive for farming without using harmful chemicals. The goals of crop rotation include:

1. **Pest and Disease Management**

Rotating crops disrupts pest and disease cycles by breaking the continuous presence of host plants, reducing the buildup of soil-borne pathogens, pests, and weeds. For example:

- **Brassica Rotation:** Alternating crops such as broccoli, cabbage, and kale with non-brassica crops like beans or tomatoes helps control cabbage worms and clubroot disease.
- **Legume Rotation:** Rotating legume crops like beans or peas with non-legume crops enhances soil fertility by fixing nitrogen and reduces the incidence of soil-borne diseases.

2. **Soil Health and Fertility**

Crop rotation improves soil health and fertility by replenishing nutrients, balancing soil pH, and reducing nutrient depletion. For example:

- **Nitrogen Fixation:** Through symbiotic partnerships with nitrogen-fixing bacteria, legume crops like peas, beans, and clover are able to fix atmospheric nitrogen. This results in the soil being enriched with this vital nutrient.
- **Cover Cropping:** Growing plants like winter rye, clover, or vetch when fields are not being used helps soil by adding natural materials, stopping weed growth, and keeping the soil in place, making it healthier overall.

3. **Weed Control**

Rotating crops disrupts weed growth cycles, reduces weed pressure, and minimizes the need for herbicides. For example:

- **Perennial Weed Management:** Alternating perennial crops with annual crops helps control perennial weeds by disrupting their growth patterns and reducing competition for resources.
- **Crop Competition:** Rotating crops with different growth habits and canopy structures effectively shades out weeds, suppresses weed germination, and reduces weed populations.

4. **Nutrient Management**

Crop rotation optimizes nutrient use efficiency by matching crop nutrient requirements with soil nutrient availability and replenishment. For example:

- **Crop-Specific Nutrient Demands:** Rotating crops with varying nutrient demands helps prevent soil nutrient imbalances and depletion, ensuring adequate levels of essential nutrients for plant growth.
- **Legume-Nonlegume Rotation:** Alternating legume and non-legume crops balances nitrogen fixation and nitrogen uptake, optimizing nitrogen availability for subsequent crops.

Planning Your Crop Rotation

1. **Assess Soil Health:** Conduct soil tests to determine nutrient levels, pH, and organic matter content, identifying soil deficiencies or imbalances that may affect crop growth.
2. **Identify Crop Groups:** Group crops into categories based on their botanical families, growth habits, and nutrient requirements to facilitate rotation planning.
3. **Plan Rotation Sequence:** Develop a rotation sequence that alternates crops from different groups to achieve pest and disease control, soil fertility, and weed suppression goals.
4. **Consider Garden Layout:** Arrange garden beds or planting areas to accommodate crop rotation sequences, ensuring adequate space for each crop and maintaining clear pathways for access and maintenance.
5. **Document Rotation Plan:** Keep detailed records of crop rotations, planting dates, and varieties grown in each bed or area to track success, identify patterns, and plan future rotations effectively.

Managing Small-Scale Agriculture

Beyond the confines of the vegetable garden, small-scale agriculture offers a broader range of opportunities for food production and self-reliance. Whether you're raising livestock, cultivating fruit trees, or growing grains and legumes, managing small-scale agriculture requires careful planning and dedication.

1. **Goal Setting:** Define clear goals and objectives for your small-scale agriculture operation. Consider factors such as production targets, income generation, sustainability goals, and personal or family needs. Setting specific, achievable, measurable, relevant, and time-bound (SMART) goals provides a roadmap for success and helps prioritize activities and resources.
2. **Site Selection and Planning:** Choose a suitable location for your small-scale agriculture operation based on factors such as soil quality, sunlight exposure, water availability, and proximity to markets or customers. Develop a comprehensive farm plan that includes layout design, crop selection, infrastructure needs, and environmental considerations. Pay attention to zoning regulations, land use restrictions, and potential environmental impacts when planning your agricultural activities.
3. **Soil Health and Fertility:** Maintaining soil health and fertility is essential for productive and sustainable agriculture. Conduct soil tests to assess nutrient levels, pH, and organic matter content, and amend soil as needed to correct deficiencies and improve soil structure. Implement practices such as crop rotation, cover cropping, composting, and organic matter application to enhance soil fertility, promote beneficial microbial activity, and prevent soil erosion and degradation.
4. **Crop Selection and Rotation:** Choose crop varieties suited to your climate, soil type, and market demand. Diversify your crop selection to spread risk, optimize resource use, and maximize market opportunities. Implement crop rotation to minimize pest and disease pressure, maintain soil fertility, and manage weed growth. Rotate crops based on their nutrient requirements, growth habits, and susceptibility to pests and diseases to achieve optimal results.
5. **Irrigation and Water Management:** Efficient irrigation and water management are crucial for small-scale agriculture, especially in regions with limited water resources. Select appropriate

irrigation methods such as drip irrigation, soaker hoses, or overhead sprinklers based on crop water needs, soil type, and water availability. Implement water-saving techniques such as mulching, rainwater harvesting, and water-efficient irrigation scheduling to conserve water, reduce runoff, and minimize water wastage.

6. **Pest and Disease Management:** Keeping bugs and diseases away from crops is really important to make sure the plants stay healthy and produce a lot of food. Implement integrated pest management (IPM) practices that combine cultural, mechanical, biological, and chemical control methods to minimize pest damage while minimizing environmental impacts. Monitor crops regularly for signs of pests and diseases, implement preventive measures such as crop rotation and companion planting, and use targeted interventions only when necessary.

7. **Harvesting and Post-Harvest Handling:** To ensure the best possible flavor, quality, and shelf life, harvest crops when they are at their peak ripeness. Take care when handling harvested food to reduce the risk of bruising, damage, and spoiling at the same time. Implement proper post-harvest handling practices such as washing, sorting, grading, and packaging to maintain freshness and quality. Store harvested produce in suitable conditions such as cool, dry, and well-ventilated spaces to prolong shelf life and minimize losses.

8. **Marketing and Sales:** Develop a marketing strategy to promote and sell your agricultural products to customers, markets, or distribution channels. Identify target markets, understand customer preferences, and differentiate your products based on quality, freshness, and uniqueness. Explore direct marketing avenues such as farmers' markets, community-supported agriculture (CSA) programs, roadside stands, and online sales platforms to connect with consumers and maximize profitability.

9. **Record Keeping and Evaluation:** Maintain accurate records of your small-scale agriculture activities, including crop planting dates, inputs, expenses, yields, and sales. Keep detailed financial records to track income and expenses, assess profitability, and make informed decisions. Evaluate your performance regularly against established goals and benchmarks, identify areas for improvement, and adjust management practices as needed to optimize productivity, efficiency, and sustainability.

Chapter 6
Livestock and Animal Husbandry

In a grid-down survival scenario, the ability to raise and care for livestock becomes paramount for long-term sustenance and resilience. Livestock not only provide valuable sources of food, such as meat, milk, and eggs, but they also offer other benefits, including manure for fertilizer, labor assistance, and potentially valuable trade commodities. However, successful animal husbandry requires careful consideration, sustainable practices, and knowledge of processing and utilization techniques. In this chapter, we will explore the essential aspects of choosing livestock, sustainable farming practices, and maximizing the use of animal products in a self-reliant lifestyle.

Choosing Livestock: Considerations and Care

When selecting livestock for a self-reliant lifestyle, several factors must be taken into account, including the climate and terrain of your location, available resources, personal preferences, and intended purposes of raising animals. Here are some key considerations:

- **Available Space:** Assess the size and layout of your property to determine the amount of space available for raising livestock. Consider factors such as pasture area, shelter requirements, and zoning regulations when selecting livestock breeds and species.
- **Resources:** Evaluate your available resources, including feed, water, fencing, and equipment, to ensure you can meet the needs of your chosen livestock. Consider the cost of purchasing and maintaining livestock, as well as ongoing expenses such as feed, veterinary care, and supplies.
- **Climate:** Choose livestock breeds and species adapted to your local climate and environmental conditions. Consider factors such as humidity, temperature extremes, rainfall, and seasonal variations when selecting animals that thrive in your region.
- **Goals:** Clarify your goals for raising livestock, whether it's producing food for your family, generating income, or providing companionship. Determine which types of livestock align with your goals and priorities, considering factors such as production potential, market demand, and personal interests.
- **Personal Preferences:** Consider your preferences regarding size, temperament, appearance, and handling requirements when choosing livestock breeds and species. Select animals that you enjoy working with and caring for, as this will enhance your overall satisfaction and success in raising livestock.

Care Practices for Raising Livestock

- **Housing and Shelter:** Provide adequate housing and shelter for your livestock to protect them from the elements, predators, and disease. Build or purchase suitable structures such as barns,

sheds, or coops designed to accommodate the specific needs of your chosen livestock breeds and species.
- **Feeding and Nutrition:** Ensure your livestock receive a balanced diet that meets their nutritional requirements for growth, reproduction, and overall health. Provide access to clean, fresh water at all times, along with appropriate feed sources such as pasture, hay, grains, and supplemental minerals as needed.
- **Health and Veterinary Care:** Maintain regular veterinary care for your livestock to prevent disease, detect health issues early, and provide necessary treatments. Develop a vaccination and deworming schedule tailored to the needs of your livestock breeds and species, and consult with a veterinarian for guidance on disease prevention and management.
- **Handling and Management:** Handle and manage your livestock with care and respect to minimize stress, injuries, and behavioral problems. Practice safe handling techniques, use appropriate equipment and facilities, and establish routines for feeding, watering, and handling to promote positive interactions and minimize risks.
- **Breeding and Reproduction:** Plan and manage breeding and reproduction activities for your livestock to maintain healthy populations and achieve your production goals. Select breeding stock with desirable traits such as productivity, conformation, and temperament, and implement breeding programs that optimize genetic diversity and performance.

Step-by-Step Projects

Building a Simple Chicken Coop

Building a chicken coop is a straightforward project that provides shelter and security for your poultry. Follow these steps to construct a basic chicken coop:

1. **Gather Materials:** Collect materials such as lumber, plywood, hardware cloth, roofing materials, screws, and nails.
2. **Design Layout:** Plan the layout and dimensions of your chicken coop, considering factors such as size, ventilation, nesting boxes, roosts, and access doors.
3. **Construct Frame:** Build the frame of the coop using pressure-treated lumber or cedar boards, assembling walls, roof, and floor sections with screws or nails.
4. **Install Hardware Cloth:** Cover openings and vents with hardware cloth to protect chickens from predators while allowing for ventilation.
5. **Add Nesting Boxes and Roosts:** Install nesting boxes and roosts inside the coop, providing comfortable and secure spaces for chickens to lay eggs and roost at night.
6. **Add Roofing and Siding:** Attach roofing materials such as corrugated metal or asphalt shingles to the coop roof, and add siding such as plywood or T1-11 to the walls for weatherproofing.
7. **Install Doors and Latches:** Install doors for access to the coop and run, adding latches or locks to secure them against predators.

8. **Add Bedding and Accessories:** Add bedding material such as straw or wood shavings to the coop floor, and provide accessories such as feeders, waterers, and dust baths for chickens' comfort and well-being.
9. **Secure Perimeter:** Fence off the perimeter of the coop and run with poultry netting or hardware cloth to prevent predators from digging under or climbing over.

Constructing a Small-Scale Goat Shelter

Providing shelter is essential for the health and well-being of goats, especially during inclement weather. Follow these steps to construct a small-scale goat shelter:

1. **Select Location:** Choose a level, well-drained area with good ventilation and access to pasture or forage.
2. **Gather Materials:** Collect materials such as pressure-treated lumber, plywood, metal roofing, screws, and nails.
3. **Design Layout:** Plan the layout and dimensions of your goat shelter, considering factors such as size, roof pitch, ventilation, and drainage.
4. **Build Frame:** Construct the frame of the shelter using pressure-treated lumber, assembling walls, roof, and floor sections with screws or nails.
5. **Install Roofing:** Attach metal roofing panels to the roof frame, ensuring proper overlap and drainage to shed water away from the shelter.
6. **Add Siding and Trim:** Install plywood or T1-11 siding to the walls of the shelter, and add trim boards or corner molding for a finished appearance.
7. **Provide Ventilation:** Cut vents or install windows with screens to provide ventilation and airflow inside the shelter, preventing moisture buildup and promoting air quality.
8. **Add Bedding and Accessories:** Spread bedding material such as straw or wood shavings on the floor of the shelter to provide insulation and comfort for goats. Provide accessories such as feeders, waterers, and mineral blocks for their nutritional needs.
9. **Secure Perimeter:** Fence off the perimeter of the shelter area with livestock fencing or panels to prevent goats from wandering off and protect them from predators.

Setting Up a Small-Scale Rabbit Hutch

Raising rabbits can be a rewarding endeavor for meat production, fur, or companionship. Follow these steps to set up a small-scale rabbit hutch:

1. **Choose Location:** Select a quiet, shaded area with good drainage and protection from wind, rain, and direct sunlight.
2. **Gather Materials:** Collect materials such as lumber, wire mesh, roofing materials, screws, and nails.
3. **Design Hutch:** Plan the design and dimensions of your rabbit hutch, considering factors such as size, ventilation, nesting boxes, and access doors.
4. **Build Frame:** Construct the frame of the hutch using pressure-treated lumber or cedar boards, assembling walls, roof, and floor sections with screws or nails.

5. **Install Wire Mesh:** Cover openings and vents with wire mesh to provide ventilation while keeping rabbits safe and secure inside the hutch.
6. **Add Roofing:** Attach roofing materials such as corrugated metal or asphalt shingles to the hutch roof, ensuring proper overlap and drainage to shed water away from the structure.
7. **Provide Nesting Boxes:** Install nesting boxes inside the hutch, providing rabbits with private, comfortable spaces for nesting and raising young.
8. **Add Bedding and Accessories:** Spread bedding material such as straw or wood shavings on the floor of the hutch to provide insulation and comfort for rabbits. Provide accessories such as feeders, waterers, and chew toys for their nutritional and behavioral needs.
9. **Secure Hutch:** Anchor the hutch securely to the ground or a stable base to prevent tipping or overturning, and install locks or latches on access doors to protect rabbits from predators.

Sustainable Animal Farming Practices

Sustainable animal farming practices aim to balance economic viability, environmental responsibility, and social equity to ensure the long-term health and well-being of both animals and ecosystems. By adopting sustainable farming practices, farmers can minimize negative impacts on the environment, conserve natural resources, and promote animal welfare while maintaining profitability and productivity.

1. **Rotational Grazing:** Implement rotational grazing systems that allow livestock to graze on diverse pastures while giving vegetation time to recover and regenerate. Rotational grazing improves soil health, reduces erosion, and enhances biodiversity by mimicking natural grazing patterns and nutrient cycling processes.
2. **Agroforestry Integration:** Integrate trees and shrubs into animal farming systems through agroforestry practices such as silvopasture, where livestock graze beneath tree canopies. Agroforestry systems provide multiple benefits, including shade for animals, improved forage quality, carbon sequestration, and enhanced wildlife habitat.
3. **Diversified Livestock:** Raise a mix of livestock species and breeds suited to local conditions and market demands, promoting genetic diversity and resilience within farming systems. Diversified livestock production reduces reliance on single species or breeds, mitigating risks associated with disease outbreaks, market fluctuations, and environmental changes.
4. **Pasture Management:** Implement sustainable pasture management practices such as rotational grazing, rest periods, and controlled stocking rates to optimize forage production, soil fertility, and animal health. Well-managed pastures enhance soil structure, water infiltration, and nutrient cycling, leading to healthier ecosystems and higher-quality forage for livestock.
5. **Integrated Pest Management (IPM):** Adopt integrated pest management (IPM) strategies to control pests and diseases in animal farming systems while minimizing reliance on synthetic pesticides and antibiotics. IPM approaches include biological controls, habitat manipulation, cultural practices, and targeted use of pesticides or pharmaceuticals as a last resort.
6. **Nutrient Cycling:** Promote nutrient cycling and resource efficiency within animal farming systems by recycling organic waste and byproducts as inputs for crop production or soil fertility

enhancement. Composting manure, bedding, and food scraps produces nutrient-rich soil amendments while decreasing dependence on chemical fertilizers and minimizing nutrient runoff.

7. **Water Conservation:** Implement water conservation practices such as rainwater harvesting, drip irrigation, and water-efficient livestock watering systems to minimize water use and protect water quality in animal farming operations. Conserving water resources reduces environmental impacts, improves resilience to drought, and enhances overall farm sustainability.

8. **Renewable Energy Integration:** Integrate renewable energy technologies such as solar panels, wind turbines, or biomass digesters into animal farming operations to reduce reliance on fossil fuels, lower greenhouse gas emissions, and offset energy costs. Renewable energy generation enhances farm resilience, diversifies income streams, and contributes to climate change mitigation efforts.

9. **Animal Welfare Standards:** Adhere to high standards of animal welfare and husbandry practices to ensure the health, comfort, and humane treatment of livestock throughout their lives. Provide animals with adequate space, shelter, nutrition, and veterinary care, and minimize stressors such as overcrowding, confinement, and rough handling.

10. **Community Engagement and Education:** Engage with local communities, consumers, and stakeholders to raise awareness about sustainable animal farming practices and promote transparency, accountability, and social responsibility within the agricultural sector. Educate consumers about the benefits of sustainable animal products and support initiatives that prioritize ethical sourcing, fair labor practices, and environmental stewardship.

Processing and Utilizing Animal Products

Once you have successfully established your livestock, the next step is efficiently processing and utilizing the various animal products they provide. This involves careful planning and resourcefulness.

Meat Processing

Meat processing involves slaughtering, butchering, and packaging livestock for consumption. Follow these steps for responsible meat processing:

- **Slaughter:** Humanely slaughter animals in compliance with animal welfare standards and regulatory requirements. Consider on-farm slaughter or local abattoirs to minimize stress and transport impacts.
- **Butchering:** Skillfully butcher carcasses into cuts of meat suitable for different culinary applications, such as steaks, roasts, and ground meat. Utilize the entire carcass to minimize waste and maximize value.
- **Packaging:** Package meat products in sanitary, labeled containers or vacuum-sealed packages to maintain freshness and quality. Follow food safety protocols to prevent contamination and ensure consumer safety.

Milk Processing

Milk processing involves pasteurizing, homogenizing, and packaging raw milk to extend shelf life and enhance safety. Consider the following steps for milk processing:

- **Pasteurization:** Heat raw milk to destroy harmful pathogens while preserving its nutritional qualities. Use low-temperature pasteurization methods when possible to retain flavor and nutrient content.
- **Homogenization:** Break down fat globules in milk to create a uniform texture and prevent cream separation. Homogenized milk has a consistent appearance and mouthfeel, enhancing consumer satisfaction.
- **Packaging:** Package pasteurized and homogenized milk in sanitary containers such as bottles, cartons, or pouches. Ensure proper refrigeration and storage to maintain freshness and prevent spoilage.

Egg Processing

Egg processing involves cleaning, grading, and packaging eggs for sale to consumers or food service establishments. Follow these steps for egg processing:

- **Cleaning:** Wash eggs with warm water and food-grade sanitizers to remove dirt, debris, and bacteria. Use gentle cleaning methods to preserve the protective cuticle and minimize eggshell damage.
- **Grading:** Sort eggs by size, weight, and quality using automated grading machines or manual inspection. Grade eggs according to industry standards such as USDA quality grades or European egg codes.
- **Packaging:** Package graded eggs in clean, labeled cartons or containers for retail sale. Include expiration dates, handling instructions, and nutritional information to inform consumers and ensure product quality.

Wool Processing

Wool processing involves shearing, washing, carding, and spinning wool fibers into yarn or fabric for textile production. Consider the following steps for wool processing:

- **Shearing:** Shear wool from sheep using electric clippers or manual shearing tools, taking care to avoid cuts or injuries to the animal. Shear sheep annually to maintain wool quality and prevent overheating.
- **Washing:** Wash raw wool in lukewarm water and mild detergents to remove grease, dirt, and lanolin. Rinse thoroughly and dry completely to prevent mold or mildew growth.
- **Carding:** Card wool fibers using hand cards, drum carders, or combing machines to align fibers and remove impurities. Carded wool can be spun into yarn or felted into fabric for various textile applications.

- **Spinning:** Spin carded wool fibers into yarn using spinning wheels, drop spindles, or spinning frames. Choose spinning techniques and yarn weights based on desired yarn characteristics and end uses.

Utilizing Byproducts

Maximize the value of animal byproducts such as bones, hides, feathers, and offal by processing them into secondary products or ingredients. Consider the following uses for animal byproducts:

- **Bones:** Boil bones to make broth or stock for cooking, or grind them into bone meal for fertilizer or animal feed.
- **Hides:** Tan hides into leather for apparel, footwear, and upholstery, or process them into gelatin for food, pharmaceutical, and industrial applications.
- **Feathers:** Clean and sterilize feathers for use in bedding, insulation, or craft projects, or process them into feather meal for animal feed or fertilizer.
- **Offal:** Render offal into fats and oils for cooking, soapmaking, or biodiesel production, or process it into pet food, fertilizer, or industrial ingredients.

Chapter 7
Foraging and Wild Edibles

Foraging refer to the practice of gathering food from wild plants and natural environments. It involves identifying and harvesting edible plants, mushrooms, fruits, nuts, and other natural resources found in forests, fields, meadows, and other habitats. Foraging can provide a diverse range of nutritious foods, including greens, berries, roots, and mushrooms, while connecting people with nature and fostering a deeper appreciation for the environment. It's important to learn how to properly identify edible species, understand their seasonal availability and habitat preferences, and practice sustainable harvesting techniques to ensure the long-term health of wild ecosystems. Foragers should also be mindful of safety considerations, such as avoiding toxic or poisonous plants and respecting wildlife habitats and conservation areas.

Identifying Edible Plants and Mushrooms

Identifying edible plants and mushrooms requires careful observation, knowledge of plant characteristics, and familiarity with local ecosystems. While there are numerous edible species found worldwide, here's a selection of 20 edible plants and 10 mushrooms commonly found in various habitats, along with tips on where to find them:

Edible Plants

- **Dandelion (Taraxacum officinale):** Found in lawns, fields, and disturbed areas, dandelion leaves are edible raw or cooked, and the flowers can be used to make tea or wine.
- **Wild Garlic (Allium vineale):** Commonly found in woodlands, meadows, and hedgerows, wild garlic leaves and bulbs have a pungent garlic flavor and can be used in salads, soups, or pesto.
- **Stinging Nettle (Urtica dioica):** Often found in damp, nutrient-rich soil, stinging nettle leaves are nutritious and can be cooked like spinach or used to make tea.
- **Plantain (Plantago major):** Widely distributed in lawns, fields, and roadsides, plantain leaves are edible raw or cooked and can be used as a salad green or medicinal herb.
- **Chickweed (Stellaria media):** Commonly found in gardens, fields, and disturbed areas, chickweed leaves and stems are edible raw or cooked and have a mild, slightly sweet flavor.
- **Wood Sorrel (Oxalis spp.):** Often found in woodlands and shady areas, wood sorrel leaves have a tart, lemony flavor and can be eaten raw or used as a garnish.
- **Burdock (Arctium spp.):** Found in fields, waste areas, and along roadsides, burdock roots, and young shoots are edible raw or cooked and have a mild, earthy flavor.
- **Wild Raspberry (Rubus spp.):** Commonly found in forests, fields, and disturbed areas, wild raspberry bushes produce sweet, juicy berries that can be eaten fresh or used in jams, jellies, or baked goods.

- **Elderberry (Sambucus spp.):** Found in hedgerows, woodlands, and riparian areas, elderberry bushes produce clusters of small, dark purple berries that can be used to make jams, syrups, or elderberry wine.
- **Wild Strawberry (Fragaria spp.):** Often found in forests, meadows, and rocky slopes, wild strawberry plants produce small, flavorful berries that can be eaten fresh or used in desserts and preserves.
- **Common Mallow (Malva neglecta):** Commonly found in gardens, fields, and disturbed areas, common mallow leaves and flowers are edible raw or cooked and have a mild, slightly mucilaginous flavor.
- **Bee Balm (Monarda spp.):** Found in woodlands, meadows, and gardens, bee balm leaves and flowers are edible raw or cooked and have a citrusy, minty flavor.
- **Purslane (Portulaca oleracea):** Often found in gardens, fields, and disturbed areas, purslane leaves and stems are edible raw or cooked and have a mild, slightly tangy flavor.
- **Violets (Viola spp.):** Commonly found in woodlands, meadows, and gardens, violet flowers and leaves are edible raw or cooked and can be used as a garnish or in salads.
- **Wild Leek (Allium tricoccum):** Found in woodlands and shaded areas, wild leek bulbs and leaves have a pungent onion flavor and can be used in soups, stews, or stir-fries.
- **Chicory (Cichorium intybus):** Often found in fields, roadsides, and waste areas, chicory leaves and roots are edible raw or cooked and have a bitter flavor that mellows when cooked.
- **Lamb's Quarters (Chenopodium album):** Commonly found in gardens, fields, and disturbed areas, lamb's quarters leaves and seeds are edible raw or cooked and have a mild, spinach-like flavor.
- **Sassafras (Sassafras albidum):** Found in woodlands and forest edges, sassafras leaves, roots, and bark are edible and can be used to make tea, root beer, or seasoning for cooking.
- **Chickweed (Stellaria media):** Commonly found in gardens, fields, and disturbed areas, chickweed leaves and stems are edible raw or cooked and have a mild, slightly sweet flavor.
- **Wild Asparagus (Asparagus officinalis):** Often found in fields, meadows, and along roadsides, wild asparagus shoots are edible and can be cooked or eaten raw.

Edible Mushrooms

- **Morel Mushroom (Morchella spp.):** Found in forests, woodlands, and disturbed areas, morel mushrooms have a distinctive honeycomb appearance and can be used in a variety of culinary dishes.
- **Chanterelle Mushroom (Cantharellus spp.):** Often found in forests, woodlands, and grassy areas, chanterelle mushrooms have a funnel-shaped cap and a delicate, fruity flavor.
- **Chicken of the Woods (Laetiporus spp.):** Found on tree trunks and logs, chicken of the woods mushrooms are bright orange or yellow and grow on trees. They have a texture like chicken meat.

- **Shaggy Mane Mushroom (Coprinus comatus):** Commonly found in grassy areas, lawns, and disturbed soil, shaggy mane mushrooms have a distinctive shaggy appearance and are best eaten when fresh.
- **Hen of the Woods (Grifola frondosa):** Often found at the base of oak trees, hen of the woods mushrooms have overlapping, fan-shaped caps and a savory, umami flavor.
- **Oyster Mushroom (Pleurotus ostreatus):** Found on dead or dying trees, oyster mushrooms have a shelf-like appearance and a delicate, seafood-like flavor.
- **Lion's Mane Mushroom (Hericium erinaceus):** Often found on hardwood trees, lion's mane mushrooms have a distinctive, shaggy appearance and a mild, seafood-like flavor.
- **Porcini Mushroom (Boletus edulis):** Found in forests and woodlands, porcini mushrooms have a distinctive brown cap with a network of white pores underneath and a rich, nutty flavor.
- **Black Trumpet Mushroom (Craterellus cornucopioides):** Often found in forests and woodlands, black trumpet mushrooms have a trumpet-shaped cap and a delicate, earthy flavor.
- **Maitake Mushroom (Grifola frondosa):** Found in forests and woodlands, maitake mushrooms have overlapping, frond-like caps and a rich, earthy flavor.

When foraging for wild edibles, it's essential to positively identify each plant or mushroom species using reliable field guides, online resources, or expert guidance from experienced foragers. Always follow sustainable harvesting practices, obtain landowner permission when foraging on private property, and prioritize safety by avoiding toxic or poisonous species and adhering to local regulations and conservation guidelines.

Ethical Foraging Practices

Ethical foraging practices are essential for maintaining the health of ecosystems, respecting wildlife habitats, and ensuring the sustainability of wild food resources. By following ethical guidelines, individuals who engage in foraging have the ability to reduce their effect on the environment, encourage biodiversity, and make a contribution to the long-term health and resilience of natural ecosystems. Here are some key principles of ethical foraging practices:

1. **Know Before You Go:** Educate yourself about local laws, regulations, and guidelines related to foraging, as well as the ecological characteristics and conservation status of target species and habitats. Obtain necessary permits or permissions when foraging on public or private lands, and respect any restrictions or prohibitions in protected areas or sensitive ecosystems.
2. **Harvest Responsibly:** Harvest wild edibles in moderation, taking only what you need and leaving behind enough plants or mushrooms to support healthy populations and reproduction. Avoid overharvesting or depleting rare, endangered, or vulnerable species, and prioritize sustainable harvesting practices that promote the regeneration and resilience of wild populations over time.
3. **Respect Wildlife Habitat:** Minimize disturbance to wildlife habitat and natural ecosystems while foraging by staying on designated trails or paths, avoiding trampling sensitive vegetation, and refraining from damaging or disrupting wildlife nests, dens, or habitats. Be mindful of the potential impacts of foraging activities on other species and ecosystems, and strive to minimize your footprint on the landscape.

4. **Practice Leave No Trace:** Leave natural areas cleaner and healthier than you found them by removing any litter, trash, or debris generated during foraging activities and disposing of waste properly. Pack out what you pack in, and avoid introducing non-native species or contaminants that could harm native wildlife or ecosystems.
5. **Identify and Respect Property Rights:** Obtain landowner permission before foraging on private property, and respect property boundaries, signage, and regulations related to foraging activities. Be aware of cultural or historical significance associated with certain plants or mushrooms and avoid harvesting from culturally sensitive or protected areas.
6. **Support Conservation Efforts:** Contribute to the conservation and protection of wild food resources and habitats by supporting local conservation organizations, participating in citizen science initiatives, and advocating for policies and practices that promote biodiversity conservation and sustainable land management. Share your knowledge and passion for foraging with others and inspire them to appreciate and protect the natural world.
7. **Cultivate Stewardship:** Cultivate a sense of stewardship and responsibility toward the natural world by fostering a deeper connection to the land, developing an understanding of ecological principles, and nurturing a sense of gratitude and reverence for the gifts of nature. Practice mindfulness and appreciation while foraging, and take time to observe and learn from the plants, mushrooms, and wildlife you encounter.

Incorporating Wild Foods into Your Diet

Wild foods offer a diverse array of flavors and nutritional benefits, making them a valuable addition to your diet. With proper preparation and creativity, you can incorporate wild edibles into your meals in delicious and inventive ways. Here are some tips for incorporating wild foods into your diet:

1. **Start with Familiar Foods:** Begin by incorporating wild foods that are similar to familiar ingredients in your diet. For example, substitute wild greens like dandelion or lambsquarters for spinach or kale in salads, soups, or stir-fries.
2. **Learn to Identify Edible Species:** Educate yourself about the edible plants, mushrooms, and other wild foods found in your region. Invest in reliable field guides, take foraging classes, or join local foraging groups to learn how to identify and harvest wild foods safely and responsibly.
3. **Forage Responsibly:** Harvest wild foods in a sustainable and ethical manner, following guidelines for responsible foraging practices. Take only what you need, leave behind enough plants or mushrooms to support healthy populations, and avoid harvesting rare, endangered, or protected species.
4. **Experiment with Recipes:** Get creative in the kitchen by experimenting with wild foods in your favorite recipes. Incorporate wild herbs, greens, or mushrooms into soups, stews, omelets, pasta dishes, or salads to add unique flavors and textures to your meals.
5. **Preserve and Store Wild Foods:** Extend the enjoyment of wild foods by preserving and storing them for later use. Dry wild herbs for seasoning, freeze berries for smoothies or baking, pickle or ferment wild vegetables for condiments or snacks, and can wild mushrooms for soups or sauces.
6. **Pair with Local Ingredients:** Combine wild foods with other locally sourced ingredients to create delicious and sustainable meals. Pair wild greens with locally grown vegetables, wild mushrooms

with locally raised meats or cheeses, and wild berries with locally produced honey or dairy products for a true taste of your region.

7. **Experiment with Flavors:** Explore the diverse flavors and textures of wild foods by experimenting with different cooking techniques and flavor combinations. Try sautéing wild greens with garlic and olive oil, roasting wild vegetables with herbs and spices, or marinating wild mushrooms in vinegar and soy sauce for a flavorful appetizer.

8. **Practice Food Safety:** Exercise caution when consuming wild foods to avoid potential risks associated with toxicity, allergies, or contamination. Always positively identify wild plants and mushrooms before consuming them, and avoid harvesting from polluted or contaminated areas.

9. **Share Your Knowledge:** Share your passion for wild foods with friends, family, and community members by introducing them to the joys of foraging and cooking with wild ingredients. Host foraging outings, cooking workshops, or potluck dinners to share your favorite wild food recipes and experiences.

PART IV
WATER MANAGEMENT

Chapter 8
Water Harvesting and Purification

In a no-grid survival scenario, access to clean water becomes a critical concern. Challenges such as contamination, scarcity, and unpredictable weather patterns amplify the significance of effective water harvesting and purification strategies. Without reliable access to clean water, the risk of waterborne illnesses, dehydration, and overall health deterioration increases significantly. Therefore, implementing sustainable water management practices and mastering purification techniques are essential for ensuring the well-being and resilience of individuals and communities.

Techniques for Collecting and Storing Water

Water collection and storage are foundational elements of any off-grid water strategy. Here are several techniques to efficiently gather and preserve this precious resource:

Rain Barrels

Rain barrels are simple and cost-effective devices used to collect rainwater from rooftops for later use. They typically consist of a large barrel or container with a lid and a spigot for accessing the collected water.

Instructions:

1. **Select a Location:** Choose a suitable location for your rain barrel near a downspout or gutter system where rainwater can be easily collected from the roof.
2. **Prepare the Barrel:** Clean and disinfect the barrel before installation to remove any dirt, debris, or contaminants. Drill a hole near the top of the barrel to accommodate the downspout.
3. **Install the Barrel:** Position the barrel underneath the downspout so that rainwater flows directly into the barrel. Use bricks or a platform to elevate the barrel slightly for easier access to the spigot.
4. **Secure the Lid:** Place the lid securely on top of the barrel to prevent debris, insects, and animals from entering the barrel and contaminating the water.
5. **Direct Overflow:** Install an overflow hose or redirect excess water away from the foundation of your home to prevent flooding or water damage.
6. **Harvest and Use Water:** Allow rainwater to collect in the barrel during rainfall events, and use the collected water for watering plants, gardening, or other non-potable uses around the home.

Cisterns

Cisterns are larger-scale water storage systems designed to collect and store rainwater for household or commercial use. They can range in size from small tanks to large underground reservoirs and are often integrated into building design or landscape features.

Instructions:
1. **Design Considerations:** Determine the size, location, and configuration of your cistern based on your water needs, available space, and local building codes and regulations.
2. **Select Materials:** Choose a suitable material for your cistern, such as concrete, plastic, or metal, based on durability, cost, and compatibility with the intended use.
3. **Prepare the Site:** Excavate a hole or trench for the cistern, ensuring that the area is level, stable, and free from obstructions. Install a stable base or foundation to support the weight of the cistern.
4. **Install the Cistern:** Lower the cistern into the prepared site using heavy machinery or manual labor. Connect the cistern to the downspouts or gutter system to collect rainwater from the roof.
5. **Secure Connections:** Ensure that all connections between the cistern, downspouts, and overflow pipes are secure and watertight to prevent leaks or water loss.
6. **Cover and Seal:** Install a secure cover or lid on the cistern to prevent debris, insects, and animals from entering the tank. Seal any openings or access points to maintain water quality and prevent contamination.
7. **Implement Filtration and Treatment:** Install filtration and treatment systems as needed to remove sediment, debris, and contaminants from the collected rainwater and ensure that it is safe for use.
8. **Monitor and Maintain:** Regularly inspect the cistern for leaks, damage, or signs of deterioration, and perform routine maintenance tasks such as cleaning filters, checking seals, and repairing any issues that arise.

Catchment Ponds

Catchment ponds, also known as rainwater harvesting ponds or reservoirs, are large surface water storage systems designed to collect and store rainwater for agricultural, industrial, or municipal use. They are typically constructed by excavating depressions in the landscape and lining them with impermeable materials to prevent seepage and retain water.

Instructions:
1. **Site Selection:** Choose a suitable location for your catchment pond based on factors such as topography, soil type, and proximity to water sources and demand centers.
2. **Design Considerations:** Determine the size, shape, and depth of the catchment pond based on your water storage needs, available land, and local hydrological conditions.
3. **Excavation and Grading:** Excavate the area for the catchment pond using heavy machinery or earthmoving equipment, taking care to create a level, uniform depression with sloping sides to maximize water storage capacity.
4. **Install Liner:** Line the excavated pond with a durable, impermeable material such as clay, geomembrane, or synthetic liner to prevent water seepage and ensure water retention.
5. **Construct Inlet and Outlet Structures:** Install inlet pipes or channels to direct rainwater runoff into the pond and outlet structures such as spillways or overflow pipes to regulate water levels and prevent flooding.

6. **Stabilize Banks:** Reinforce the banks of the catchment pond with riprap, vegetation, or other erosion control measures to prevent erosion, sedimentation, and loss of water storage capacity.
7. **Implement Filtration and Treatment:** Install sediment traps, baffles, or filtration systems to take out sediment, debris, and contaminants from the collected rainwater before it enters the pond.
8. **Monitor and Maintain:** Regularly monitor water levels, inspect the integrity of the pond liner, and perform routine maintenance tasks such as cleaning inlet and outlet structures, removing debris, and repairing any damage to ensure the long-term functionality and effectiveness of the catchment pond.

DIY Filtration and Purification Methods

While collecting water is crucial, ensuring its cleanliness is equally important. DIY filtration and purification methods offer practical solutions to make water safe for consumption. These methods can remove impurities, sediment, pathogens, and contaminants from water, making it clean and potable. Here are some simple and effective DIY filtration and purification methods:

Boiling

Heating water to its boiling point stands as one of the easiest and most efficient means of purifying it. This process effectively eliminates viruses, bacteria, and other harmful pathogens that might inhabit the water.

Steps:

1. Bring the water to a rolling boil in a clean pot or kettle.
2. Allow the water to boil for almost 1 minute (or 3 minutes at higher altitudes) to ensure that all pathogens are killed.
3. Let the water cool before drinking or storing it.

Cloth Filtration

Cloth filtration is a basic method for removing sediment and larger particles from water. It can be done using a clean cloth or fabric.

Steps:

1. Place a clean cloth or fabric over the mouth of a container.
2. Slowly pour the water through the cloth, allowing it to filter out sediment and particles.
3. Repeat the process if necessary until the water appears clear and free from visible impurities.

Sand and Gravel Filtration

Sand and gravel filtration is a simple and effective method for removing sediment, debris, and some bacteria from water.

Steps:

1. Create a filtration system using layers of coarse gravel, fine gravel, and sand in a container such as a plastic bucket or PVC pipe.
2. Pour the water through the layers of gravel and sand, allowing it to pass through the filtration medium.

3. Gather the filtered water in a clean container for drinking or storage.

Charcoal Filtration

Charcoal filtration is a method for removing impurities, odors, and some chemicals from water. Activated charcoal is particularly effective for this purpose.

Steps:
1. Fill a container with activated charcoal or crushed charcoal pieces.
2. Pour the water through the charcoal, allowing it to filter out impurities and contaminants.
3. Gather the filtered water in a clean container for drinking or storage.

Solar Disinfection (SODIS)

Solar disinfection is a natural purification method that uses sunlight to kill bacteria, viruses, and other pathogens in water.

Steps:
1. Fill a clear plastic bottle with water from a clean source.
2. Bring the bottle in direct sunlight for at least six hours (or longer on cloudy days) to allow the UV radiation from the sun to disinfect the water.
3. After solar disinfection, the water may still contain sediment and particles, so it may be necessary to filter it through cloth or charcoal before drinking.

DIY Ceramic Filter

A DIY ceramic filter can effectively remove bacteria and sediment from water. It involves using porous ceramic material as a filter medium.

Steps:
1. Obtain a ceramic pot or container with small pores.
2. Fill the container with water and allow it to seep through the pores, trapping bacteria and sediment.
3. Gather the filtered water in a clean container for drinking or storage.

DIY Berkey Water Filter

A DIY Berkey water filter can effectively remove impurities, pathogens, and contaminants from water using a combination of charcoal and other filtration media.

Steps:
1. Obtain two clean food-grade buckets, with one smaller than the other.
2. Drill holes in the bottom of the smaller bucket then place it inside the larger bucket.
3. Fill the smaller bucket with layers of charcoal, sand, and gravel, with a cloth filter at the bottom.
4. Pour the water into the smaller bucket and allow it to filter through the layers of media.
5. Collect the filtered water in the larger bucket for drinking or storage.

Building and Maintaining Water Systems

Building and maintaining water systems is crucial for ensuring access to clean, safe, and reliable water for various purposes, including sanitation, drinking, agriculture, and industry. Properly designed and maintained water systems can enhance water security, promote public health, and support sustainable development. Here's a guide to building and maintaining water systems:

Planning and Design

- Assess Water Needs: Determine the water requirements for the intended use (e.g., household, agricultural, industrial) and estimate the demand based on population size, consumption patterns, and seasonal variations.
- Identify Water Sources: Identify potential water sources such as groundwater, surface water, rainwater, or treated wastewater, and evaluate their availability, quality, and sustainability.
- Choose Appropriate Technologies: Select appropriate technologies and components for water collection, storage, treatment, and distribution based on site-specific conditions, budget constraints, and local regulations.
- Design System Layout: Develop a comprehensive layout and schematic diagram for the water system, including pipelines, pumps, storage tanks, treatment facilities, and distribution networks.

Construction and Installation

- Excavation and Site Preparation: Prepare the site for construction by clearing vegetation, leveling the ground, and excavating trenches for pipelines, storage tanks, and other infrastructure.
- Install Components: Install water collection systems (e.g., rainwater harvesting systems), storage tanks, treatment units (e.g., filtration, disinfection), pumps, valves, meters, and distribution pipelines according to the design specifications and manufacturer's instructions.
- Ensure Quality and Safety: Use high-quality materials, equipment, and construction techniques to ensure the durability, reliability, and safety of the water system. Adhere to relevant building codes, standards, and regulations to meet quality and safety requirements.

Testing and Commissioning

- Conduct Performance Testing: Test the functionality, efficiency, and performance of each component and subsystem of the water system, including water quality parameters, flow rates, pressure levels, and system operation.
- Address Deficiencies: Identify and address any deficiencies, leaks, or malfunctions in the water system through repairs, adjustments, or replacements as needed to ensure optimal performance and reliability.
- Commission System: Commission the water system by systematically activating and testing each component and subsystem to ensure proper functioning and integration. Document the commissioning process and results for future reference.

Operation and Maintenance

- Develop Maintenance Plan: Develop a comprehensive maintenance plan outlining routine inspection, servicing, and repair tasks for each component of the water system. Schedule regular maintenance activities based on manufacturer recommendations, operational requirements, and environmental conditions.

- Monitor Water Quality: Regularly check the water to make sure it's clean and safe by looking at things like how acidic it is, how clear it is, how much chlorine is in it, and if there are any harmful germs in it. Conduct periodic water testing and analysis to identify potential issues or emerging contaminants.

- Address Repairs and Upgrades: Promptly address any repairs, replacements, or upgrades needed to maintain the functionality, efficiency, and safety of the water system. Keep records of maintenance activities, repairs, and upgrades for documentation and analysis.

- Train Personnel: Provide training and capacity-building opportunities for personnel responsible for operating and maintaining the water system. Ensure that staff are knowledgeable about system operation, maintenance procedures, safety protocols, and emergency response measures.

Community Engagement and Participation

- Foster Community Engagement: Engage stakeholders, including NGOs, local communities, government agencies, and private sector partners, in the planning, design, implementation, and management of water systems. Encourage community involvement, participation, and ownership to build resilience, promote sustainability, and foster social cohesion.

- Raise Awareness: Increase the level of understanding among the members of the community on the significance of water conservation, sanitary habits, and the management of water in a sustainable manner. Provide education and outreach initiatives on water-related issues, including water quality, sanitation, and environmental stewardship.

- Promote Equity and Inclusivity: Ensure equitable access to water services for all members of the community, comprising marginalized groups, women, children, and people with disabilities. Design water systems with a focus on inclusivity, affordability, and accessibility to address the diverse needs and priorities of the population.

Chapter 9

Sustainable Water Use

As water scarcity becomes an increasing global concern, adopting sustainable water use practices is not only responsible but imperative. In a no-grid survival scenario, efficient water management ensures the longevity of your water supply. This chapter explores key aspects of sustainable water use, including conserving water in daily activities, implementing greywater and rainwater systems, and managing water in various climates.

Conserving Water in Daily Activities

Conserving water is an essential component of sustainable living, especially in a survival setting where resources are limited. Implementing water-saving practices in your daily activities can significantly extend your available water supply.

1. **Collect Rainwater:** Utilize rain barrels, buckets, or tarps to collect rainwater from rooftops or other surfaces for drinking, cooking, and hygiene purposes.
2. **Reuse Graywater:** Reuse wastewater from activities such as dishwashing, laundry, and bathing for flushing toilets, watering plants, or cleaning surfaces.
3. **Take Shorter Showers:** Limit shower time and use a bucket to capture water while waiting for it to warm up, then repurpose this water for other tasks like flushing toilets or watering plants.
4. **Fix Leaks:** Regularly inspect and repair leaks in plumbing fixtures, pipes, and irrigation systems to prevent water loss and conserve valuable resources.
5. **Turn Off Taps:** Turn off taps tightly to avoid drips and leaks, and only run water when necessary, such as while brushing teeth or washing dishes.
6. **Use Efficient Fixtures:** Install water-saving tools like toilets, faucets, and showerheads that use less water but still work well.
7. **Sweep, Don't Hose:** Use a broom or rake instead of a hose to clean outdoor surfaces like driveways, sidewalks, and decks to conserve water.
8. **Water Wisely:** By watering outdoor plants and gardens early in the morning or late in the evening, you can reduce the amount of water that is lost to evaporation and increase the amount that is absorbed. Additionally, you can use drip irrigation or soaker hoses to feed water straight to the roots of the plants.
9. **Mulch and Compost:** Mulch garden beds and compost organic materials to retain moisture in the soil, reduce evaporation, and improve soil health, reducing the need for irrigation.
10. **Cook Smart:** Use minimal water for cooking by steaming or microwaving food instead of boiling, and reuse cooking water for soups, stews, or watering plants after it cools.
11. **Opt for Efficient Cleaning:** Use a basin or stopper in the sink when washing dishes to minimize water usage, and scrape dishes instead of rinsing them before loading the dishwasher.

12. **Educate and Involve Others:** Raise awareness about the importance of water conservation and encourage friends, family, and community members to adopt water-saving practices in their daily lives.

Implementing Greywater and Rainwater Systems

Harnessing alternative water sources like greywater and rainwater expands your water supply capacity. By understanding how to collect and utilize these sources effectively, you can enhance your self-reliance and reduce dependence on traditional water supplies.

Greywater System

Greywater systems capture and treat wastewater from household activities to reuse it for irrigation, toilet flushing, and other non-potable purposes, reducing the demand for fresh water and minimizing wastewater discharge.

Step-by-Step Procedure:
1. **Assess Feasibility:** Determine if your home and local regulations permit the installation of a greywater system. Assess the feasibility of retrofitting existing plumbing or incorporating greywater collection into new construction.
2. **Select Source:** Identify sources of greywater, such as sinks, showers, baths, and washing machines, and plan for their diversion and collection.
3. **Diversion and Filtration:** Install diversion valves or pipes to reroute greywater from fixtures to a collection point. Install filters or screens to remove large particles and debris from the greywater to prevent clogs and blockages.
4. **Treatment:** Depending on local regulations and system design, treat greywater using methods such as filtration, settling, disinfection, or biological processes to remove contaminants and pathogens and improve water quality.
5. **Storage:** Store treated greywater in a dedicated tank or cistern for later use in irrigation or toilet flushing. Ensure proper sizing and sealing of the storage tank to prevent leaks, contamination, and odors.
6. **Distribution:** Install separate plumbing lines or distribution systems to deliver treated greywater to irrigation zones, toilet cisterns, or other non-potable fixtures throughout the property.
7. **Maintenance:** Regularly inspect and maintain the greywater system, including cleaning filters, monitoring water quality, and addressing any leaks or malfunctions promptly. Follow manufacturer recommendations and local regulations for maintenance and operation.

Rainwater Harvesting System

Those systems that collect rainwater from rooftops or other surfaces and store it for later utilization in drinking water are known as rainwater harvesting systems, irrigation, and other domestic applications, reducing reliance on municipal water supplies and groundwater sources.

Step-by-Step Procedure:

1. **Assess Roof Area:** Determine the size and slope of the roof area available for rainwater collection. Calculate the potential volume of rainwater that can be harvested based on annual precipitation and roof dimensions.
2. **Gutter Installation:** Install gutters and downspouts along the roof edges to capture rainwater runoff and direct it to a central collection point or storage tank. Ensure proper sizing, slope, and alignment of gutters for efficient drainage.
3. **Filtration and Pre-Treatment:** Install leaf guards, screens, or mesh filters at gutter outlets to remove debris, leaves, and other contaminants from the rainwater before it enters the storage tank. Consider additional pre-treatment measures such as first flush diverters or sediment traps to improve water quality.
4. **Storage Tank:** Select and install a suitable storage tank or cistern to store harvested rainwater. Choose a location that is accessible, level, and stable, and ensure proper sizing and sealing of the tank to prevent leaks, contamination, and mosquito breeding.
5. **Overflow and Outlet:** Install overflow pipes or outlets to divert excess rainwater away from buildings and foundations to prevent flooding or water damage. Connect outlet pipes to distribution systems or irrigation networks for controlled use of harvested rainwater.
6. **Water Treatment:** Depending on intended uses and water quality requirements, consider treating harvested rainwater with filtration, disinfection, or UV sterilization to remove contaminants and pathogens and ensure water safety.
7. **Maintenance:** Regularly inspect and maintain the rainwater harvesting system, including cleaning gutters, screens, and filters, checking tank levels, and repairing any leaks or damage. Monitor water quality and perform routine maintenance tasks according to manufacturer recommendations and local regulations.

Managing Water in Various Climates

Survival scenarios can unfold in diverse climates, each presenting unique challenges and opportunities for water management. Understanding how to adapt water practices to different climates ensures flexibility and resilience in your self-reliance efforts.

Arid and Semi-Arid Climates

- Emphasize Water Conservation: Prioritize water conservation practices such as rainwater harvesting, greywater recycling, and efficient irrigation techniques like drip irrigation to minimize water loss.
- Implement Xeriscaping: Design landscapes using drought-tolerant native plants, mulch, and soil amendments to reduce water usage and promote soil moisture retention.
- Maximize Shade and Evaporation Control: Strategically place trees and shrubs to offer shade and minimize soil evaporation, while employing shading structures such as pergolas or shade cloth to safeguard delicate plants and preserve moisture.

- Store and Treat Water: Collect and store rainwater in cisterns or tanks for later use, and implement water treatment methods such as filtration and disinfection to ensure water safety and quality.

Tropical and Humid Climates

- Manage Excess Rainfall: Implement drainage systems, swales, and rain gardens to manage excess rainfall and prevent flooding or soil erosion, and divert runoff to storage tanks or infiltration basins for groundwater recharge.
- Utilize Evapotranspiration: Leverage natural processes like evaporation and transpiration to regulate soil moisture levels, and select plant species that can tolerate high humidity and fluctuating moisture conditions.
- Monitor Water Quality: Regularly test and monitor water quality in rivers, streams, and groundwater sources for contamination and pollution, and implement water treatment measures as needed to ensure safe drinking water.
- Prevent Waterborne Diseases: Educate communities about waterborne diseases and promote hygiene practices such as proper handwashing and sanitation to prevent the spread of illnesses in humid climates where microbial growth is prevalent.

Temperate and Mediterranean Climates

- Optimize Rainwater Harvesting: Capture and store rainwater from rooftops and other surfaces in barrels, cisterns, or ponds for use during dry periods or for irrigation, and utilize permeable surfaces and rain gardens to enhance groundwater recharge.
- Manage Seasonal Variation: Plan for seasonal variations in precipitation and water availability by storing excess water during wet periods and implementing water-saving measures during dry spells, such as mulching, soil moisture monitoring, and efficient irrigation scheduling.
- Promote Soil Health: Implement soil conservation and management practices such as composting, cover cropping, and rotational grazing to enhance soil structure, water retention, and nutrient cycling in temperate and Mediterranean climates.
- Diversify Water Sources: Search alternative water sources such as recycled wastewater, greywater, or desalinated seawater to supplement conventional water supplies and reduce reliance on limited freshwater resources.

Mountainous and Alpine Climates

- Manage Snowmelt Runoff: Implement strategies to capture and store snowmelt runoff in reservoirs, ponds, or infiltration basins for use during dry seasons or for hydropower generation, and monitor snowpack levels and melt rates to forecast water availability.
- Protect Watersheds: Preserve and protect mountain ecosystems and watersheds through land conservation, reforestation, and erosion control measures to maintain water quality and quantity in alpine environments.

- Adapt to Glacial Retreat: Prepare for changes in water availability and glacier melt rates resulting from climate change by developing adaptive strategies such as water storage, diversification of water sources, and community resilience planning in mountainous regions.

Regardless of the climate, effective water management in a no-grid survival situation requires a combination of conservation, storage, treatment, and sustainable use practices tailored to local conditions and resources. By implementing these strategies, individuals and communities can enhance their resilience and self-sufficiency in water management, even in challenging environmental circumstances.

PART V

SHELTER AND HOMEBUILDING

Chapter 10
Designing Your Off-Grid Home

Designing an off-grid home presents unique challenges and opportunities. From selecting the right location and materials to implementing efficient building techniques and customizing your living space, every aspect requires careful consideration. In a no-grid scenario, your home becomes the foundation of self-reliance, providing shelter, comfort, and security. This chapter explores the essential elements of designing your off-grid home, addressing the challenges and significance of each aspect in creating a sustainable and resilient living environment.

Choosing the Right Location and Materials

Selecting the optimal location and materials for your off-grid home lays the foundation for a sustainable and resilient living space. Careful consideration of environmental factors and material choices is crucial.

Choosing the Right Location

Selecting the right location for your off-grid home is crucial for maximizing access to natural resources and minimizing environmental impact. Take into account the following considerations when selecting a location:

1. **Access to Sunlight:** Opt for a location with ample sunlight exposure, especially if you plan to rely on solar energy for power generation. South-facing slopes or open areas without obstruction from trees or buildings are ideal for installing solar panels.
2. **Water Availability:** Ensure access to water sources for drinking, irrigation, and other domestic needs. Look for sites near rivers, streams, lakes, or groundwater aquifers, or when it comes to collecting and storing rainwater, you should think about installing rainwater harvesting systems.
3. **Climate Considerations:** Take into account local climate conditions such as temperature, precipitation, wind patterns, and seasonal variations. Choose a location that offers natural protection from extreme weather events and harsh environmental conditions.
4. **Soil Quality:** Assess soil quality and suitability for construction, gardening, and landscaping. Conduct soil tests to determine soil composition, drainage characteristics, and fertility levels to inform site preparation and landscaping plans.
5. **Legal and Regulatory Considerations:** Ensure compliance with local zoning regulations, building codes, environmental regulations, and land use restrictions when selecting a site for your off-grid home. Ensure that you acquire all required permits and approvals prior to commencing construction to prevent any potential legal issues.
6. **Accessibility:** Consider accessibility and transportation infrastructure when choosing a location for your off-grid home. Ensure access to roads, trails, or other transportation routes for commuting, emergency services, and logistical purposes.

Choosing Sustainable Materials

Selecting sustainable materials for your off-grid home can minimize environmental impact, promote energy efficiency, and enhance durability and resilience. Consider the following factors when choosing construction materials:

1. **Renewable and Recycled Materials:** Prioritize materials that are renewable, recycled, or locally sourced to reduce embodied energy and carbon footprint. Choose materials like sustainably harvested wood, bamboo, recycled metal, reclaimed timber, or salvaged materials from demolition sites.

2. **Energy Efficiency:** Select materials with high thermal performance and energy efficiency to minimize heat loss, improve insulation, and reduce heating and cooling costs. Consider options like insulated concrete forms (ICFs), structural insulated panels (SIPs), or natural insulation materials like straw bales, cellulose, or wool.

3. **Durability and Longevity:** Choose materials that are durable, weather-resistant, and resistant to pests, decay, and degradation. Invest in high-quality materials like treated lumber, metal roofing, fiber cement siding, or durable finishes and sealants to ensure the longevity of your off-grid home.

4. **Low-Environmental Impact:** Consider the environmental impact of materials throughout their lifecycle, including extraction, production, transportation, installation, use, and disposal. Choose materials with low embodied energy, minimal greenhouse gas emissions, and minimal environmental pollution.

5. **Non-Toxic and Healthy:** Prioritize materials that are non-toxic, low-VOC (volatile organic compound), and free from harmful chemicals and additives to ensure indoor air quality and occupant health. Choose natural and eco-friendly materials like clay, lime, natural paints, and finishes.

6. **Adaptability and Repairability:** Select materials that are adaptable, modular, and easy to repair or replace to accommodate future changes, expansions, or renovations. Choose materials with long-term availability and compatibility with local construction practices and traditions.

Building Techniques for Durability and Efficiency

The construction techniques employed in building your off-grid home significantly impact its durability, energy efficiency, and overall resilience. Utilizing appropriate building techniques ensures a structure that can withstand the challenges of a no-grid environment. Here are some key techniques to consider:

1. **Passive Solar Design:** To get the most out of natural heating and cooling, include passive solar design principles into your project. Orient the home to capture sunlight during the winter months for warmth and use shading devices like overhangs or deciduous trees to block excessive heat gain in the summer.

2. **Insulation:** Make an investment in insulation of a high grade to cut down on heat loss during the winter and heat gain during the summer. Use materials like fiberglass, cellulose, foam, or natural insulation such as straw bales to create a thermal barrier and improve energy efficiency.

3. **Thermal Mass:** Incorporate thermal mass materials such as concrete, brick, or stone into the construction of the house. These materials effectively capture and retain heat during the day,

gradually releasing it at night. This process aids in moderating indoor temperatures, thereby diminishing the need for excessive heating and cooling.

4. **Air Sealing:** Ensure proper air sealing to minimize air infiltration and heat loss. Seal gaps, cracks, and penetrations in the building envelope with caulking, weatherstripping, or spray foam insulation to create a tight thermal envelope and improve energy efficiency.
5. **Energy-Efficient Windows and Doors:** Install energy-efficient windows and doors with low-E coatings, multiple glazing layers, and insulated frames to reduce heat transfer and improve thermal performance. Choose windows and doors with appropriate U-values and solar heat gain coefficients for your climate.
6. **Natural Ventilation:** Design the home to facilitate natural ventilation and airflow to reduce the need for mechanical cooling. Incorporate operable windows, clerestory windows, or skylights to encourage cross-ventilation and passive cooling through natural convection.
7. **Rainwater Management:** Implement rainwater management strategies to control runoff, prevent erosion, and capture rainwater for onsite use. Install gutters, downspouts, and rainwater harvesting systems to collect and store rainwater for irrigation, toilet flushing, and other non-potable uses.
8. **Durable Exterior Finishes:** Choose durable exterior finishes and cladding materials that can withstand weathering, UV exposure, and moisture. Options include fiber cement siding, metal roofing, stone veneer, or naturally weather-resistant wood species like cedar or redwood.
9. **Foundation and Structural Design:** Design the foundation and structural system to withstand seismic activity, high winds, and other environmental hazards. Consider reinforced concrete foundations, steel framing, or engineered wood products to enhance structural integrity and resilience.
10. **Low-Maintenance Landscaping:** Create low-maintenance landscaping features with native plants, drought-tolerant species, and permeable paving materials to minimize water usage, reduce landscaping maintenance, and enhance curb appeal.
11. **Off-Grid Utilities:** Optimize off-grid utilities such as solar power, wind turbines, micro-hydro systems, or composting toilets to provide self-sufficient and sustainable energy and water solutions. Set up energy-saving appliances and lighting fixtures to reduce energy usage and optimize efficiency.
12. **Modular and Prefabricated Construction:** Consider modular or prefabricated construction methods to streamline the building process, reduce construction waste, and achieve consistent quality control. Prefabricated panels, modules, or components can be assembled onsite quickly and efficiently, saving time and labor costs.

Customizing Your Living Space

Designing your living space within an off-grid home enables you to craft a cozy, practical, and tailored environment tailored to suit your individual requirements and tastes. Here's a guide on how to create a personalized and well-crafted living environment:

1. **Functional Layout:** Design the layout of your living space to optimize functionality and flow. Consider how you will use each room and arrange furniture, appliances, and storage to maximize usability and efficiency.
2. **Multi-Purpose Spaces:** Create multi-purpose spaces that serve multiple functions to maximize flexibility and utility. For example, a guest room could double as a home office or hobby space, or a dining area could also function as a workspace or study area.
3. **Storage Solutions:** Incorporate ample storage solutions to keep your living space organized and clutter-free. Utilize built-in shelving, cabinets, closets, and under-bed storage to maximize storage capacity while minimizing visual clutter.
4. **Comfortable Furnishings:** Choose comfortable and ergonomic furnishings that enhance comfort and well-being. Invest in quality seating, bedding, and lighting to create inviting and cozy living spaces where you can relax and unwind.
5. **Personal Touches:** Infuse your living space with personal touches and decor that reflect your personality, interests, and style preferences. Display artwork, photos, souvenirs, and mementos that hold sentimental value and contribute to a sense of home.
6. **Natural Elements:** Integrate elements of nature into your interior design to bring the outside inside like wood, stone, plants, and natural fibers into your living space. Use natural materials for flooring, furniture, and decor to create a warm and inviting atmosphere.
7. **Energy-Efficient Lighting:** Install energy-efficient lighting fixtures and bulbs to minimize energy consumption and reduce reliance on off-grid power sources. Opt for LED or CFL bulbs, solar-powered lighting, or daylighting strategies to maximize natural light and minimize electricity usage.
8. **Flexible Design:** Design your living space with flexibility in mind to accommodate changing needs and preferences over time. Choose modular furniture, movable partitions, and adaptable layouts that can be easily reconfigured or expanded as your lifestyle evolves.
9. **Green Building Materials:** Incorporate sustainable and environmentally friendly building materials into your living space to minimize environmental impact and promote sustainability. Choose materials with low embodied energy, recycled content, and eco-friendly certifications.
10. **Outdoor Living Spaces:** Extend your living space outdoors with functional and inviting outdoor areas. Create outdoor seating areas, dining areas, gardens, and recreational spaces where you can connect with nature and enjoy the beauty of your surroundings.
11. **Off-Grid Amenities:** Customize your living space with off-grid amenities and features that enhance self-sufficiency and sustainability. Consider installing rainwater harvesting systems, composting toilets, renewable energy systems, and other off-grid utilities to minimize reliance on external resources.
12. **Accessibility:** Ensure that your living space is accessible and inclusive for people of all ages and abilities. Incorporate universal design principles, such as wide doorways, zero-step entries, and accessible bathrooms, to accommodate individuals with mobility challenges.

Chapter 11
Energy-Efficient Home Solutions

In a world increasingly affected by climate change and resource depletion, energy efficiency has become a crucial aspect of sustainable living. Designing and implementing energy-efficient solutions in your off-grid home not only reduces environmental impact but also lowers energy costs and enhances resilience.

Insulation and Passive Solar Design

Insulation and passive solar design are two key elements of energy-efficient home construction, particularly in off-grid or no-grid settings where minimizing energy consumption is essential for sustainability and self-sufficiency.

Insulation

Insulation is vital for controlling indoor temperatures, as it diminishes the exchange of heat between a building's interior and exterior. In colder regions, insulation conserves warmth within homes throughout winter, whereas in warmer areas, it prevents heat infiltration from outside, keeping the indoors cool. By establishing a thermal barrier, efficient insulation diminishes reliance on heating and cooling systems, leading to decreased energy usage and lower utility expenses.

There are some types of insulation materials commonly used in residential construction, including fiberglass, cellulose, foam board, and spray foam. Each type has its own characteristics and advantages, such as R-value (thermal resistance), fire resistance, moisture resistance, and installation method. When selecting insulation materials, it's important to consider factors such as building design, climate, budget, and environmental impact.

Proper installation of insulation is crucial for maximizing its effectiveness. Insulation should be installed uniformly and securely throughout the building envelope, including walls, floors, ceilings, and attics, to minimize thermal bridging and air leakage. Sealing gaps, cracks, and penetrations with caulking, weatherstripping, or spray foam insulation helps prevent air infiltration and ensures a tight thermal envelope.

In addition to traditional insulation materials, natural and eco-friendly alternatives such as wool, cotton, cellulose, and straw bales are gaining popularity for their sustainability and low environmental impact. These materials offer excellent thermal performance and can be sourced locally, reducing embodied energy and carbon footprint.

Passive Solar Design

Passive solar design harnesses the sun's energy to heat and cool buildings naturally, without the need for mechanical heating or cooling systems. By strategically orienting and designing a home to capture and utilize solar energy, passive solar design maximizes comfort, energy efficiency, and sustainability.

Key principles of passive solar design include:

1. **Orientation:** Orienting the home's long axis within 15 degrees of true south maximizes exposure to the sun's path throughout the day, allowing for optimal solar gain in winter and minimizing heat gain in summer.
2. **Glazing:** Installing south-facing windows with a high solar heat gain coefficient (SHGC) and appropriate overhangs or shading devices to control solar heat gain and maximize natural daylighting.
3. **Thermal Mass:** Utilizing thermal mass materials like concrete, brick, or tile within the interior of the house to capture and retain solar heat throughout the day, subsequently releasing it slowly during the night, thereby assisting in maintaining consistent indoor temperatures.
4. **Insulation:** Providing adequate insulation and air sealing to minimize heat loss and maintain thermal comfort, particularly in colder climates where passive solar heating is most beneficial.
5. **Natural Ventilation:** Designing for cross-ventilation and natural airflow to facilitate passive cooling in warmer months, using operable windows, vents, and clerestory openings to promote fresh air circulation and thermal comfort.
6. **Overhangs and Shading:** Using overhangs, awnings, or deciduous trees to provide shade and block direct sunlight during the hottest times of the day, reducing the need for mechanical cooling and minimizing solar heat gain.

Renewable Energy Options for Home Use

Renewable energy options offer sustainable and environmentally friendly alternatives for powering homes, particularly in off-grid or no-grid settings where access to traditional utility services may be limited. Here are some common renewable energy options for home use:

- **Solar Power:** Solar power stands out as a leading choice among renewable energy options for home use due to its widespread availability and popularity. Utilizing semiconductor materials, solar panels, or photovoltaic (PV) panels, transform sunlight into electricity. These panels can be affixed on roofs, set up in ground-mounted arrays, or seamlessly integrated into building facades, providing a sustainable source of electricity to power various household needs such as lighting, appliances, heating systems, and more. Additionally, solar water heating systems use sunlight to heat water for domestic use, such as showers, laundry, and dishwashing.
- **Wind Power:** Utilizing the kinetic energy of moving air, wind energy is converted into electricity via wind turbines. These turbines can be implemented on residential properties in smaller scales to either complement or reduce reliance on grid electricity. Optimal performance is observed in regions with steady and robust wind speeds, such as coastal areas or expansive plains. Proper siting and installation are crucial to maximize energy production and minimize noise and visual impacts.
- **Hydroelectric Power:** Hydroelectric power utilizes the energy of flowing water, such as rivers or streams, to generate electricity through turbines. Micro-hydro systems can be installed on properties with access to flowing water sources to generate renewable energy for off-grid homes.

Micro-hydro systems require careful planning and design to ensure sufficient water flow and head height for efficient electricity generation.
- **Biomass Energy:** Biomass energy involves the combustion or conversion of organic materials such as wood, agricultural residues, or biofuels to produce heat or electricity. Wood-burning stoves, pellet stoves, and biomass boilers can provide space heating and hot water for residential use. Additionally, biogas digesters can convert organic waste materials into methane gas for cooking, heating, or electricity generation.
- **Geothermal Energy:** Geothermal energy utilizes the natural heat stored beneath the Earth's surface to provide space heating, hot water, and electricity. Geothermal heat pumps harness heat from the earth during winter to warm buildings and return heat to the ground in summer for cooling. While these systems are eco-friendly and efficient, initial installation expenses can be significant.
- **Hybrid Systems:** Optimum energy output and dependability can be achieved through the utilization of hybrid renewable energy systems, which mix various renewable energy sources. For example, a hybrid system may combine solar panels with wind turbines or a backup generator to ensure continuous power supply in varying weather conditions or seasonal changes. Hybrid systems offer increased energy independence and resilience compared to single-source systems.

When considering renewable energy options for home use, it's important to conduct a thorough assessment of energy needs, site conditions, available resources, and budget constraints. Consulting with renewable energy professionals, engineers, and energy auditors can help homeowners evaluate the feasibility and potential benefits of different renewable energy technologies for their specific needs and circumstances.

Maintaining and Upgrading Your Home

Ensuring regular maintenance and upgrades is crucial to enhance energy efficiency and extend the lifespan of your off-grid residence. Here's how to keep your home running smoothly:

Regular Maintenance

1. Conduct regular inspections of your home's exterior and interior to identify any signs of damage, wear, or deterioration.
2. Perform routine maintenance tasks such as trimming vegetation, cleaning gutters, and repairing roof leaks to prevent water damage and structural issues.
3. Check and replace weatherstripping, caulking, and seals around windows and doors to prevent air leaks and improve energy efficiency.
4. Service and maintain off-grid utilities such as solar panels, wind turbines, batteries, and water systems to ensure optimal performance and longevity.
5. Regularly inspect and maintain HVAC systems to enhance indoor air quality and optimize energy efficiency.
6. Ensure the safety and compliance of your home by regularly testing and servicing smoke detectors, carbon monoxide detectors, and fire extinguishers.

Upgrades and Improvements
1. Think about upgrading to energy-efficient appliances, lighting fixtures, and HVAC systems as a means to cut down on energy usage and decrease utility expenses.
2. Install water-saving fixtures such as showerheads, low-flow toilets, and faucets to conserve water and minimize wastewater generation.
3. Explore renewable energy options such as additional solar panels, wind turbines, or micro-hydro systems to increase energy independence and resilience.
4. Upgrade insulation, windows, and doors to improve thermal performance and enhance comfort while reducing heating and cooling loads.
5. Invest in energy storage systems such as battery banks or backup generators to ensure continuous power supply during outages or periods of low renewable energy production.
6. Implement smart home technologies such as programmable thermostats, remote monitoring systems, and home automation devices to optimize energy usage and control home systems remotely.

Safety and Security
1. Enhance home security measures with motion-sensor lights, surveillance cameras, and alarm systems to deter intruders and protect property.
2. Install carbon monoxide detectors and radon gas detectors to monitor indoor air quality and mitigate health risks.
3. Secure loose items and outdoor furniture during severe weather events to prevent damage and injury from high winds and flying debris.
4. Develop emergency preparedness plans and stockpile essential supplies such as food, water, first aid kits, and emergency lighting in case of prolonged power outages or natural disasters.
5. Practice fire safety measures such as regular chimney cleaning, proper storage of flammable materials, and having fire extinguishers readily available in strategic locations throughout the home.

PART VI
POWER INDEPENDENCE

Chapter 12
Solar Power Solutions

In the pursuit of off-grid living, solar power stands out as a primary and sustainable energy solution. Harnessing the sun's energy allows for independence from traditional power grids, fostering resilience and environmental responsibility. This chapter delves into the basics of solar energy and panel installation, the crucial aspects of battery storage and power management, and strategies for maximizing efficiency in solar power systems.

Basics of Solar Energy and Panel Installation

Solar power is an enduring and eco-friendly energy solution that utilizes sunlight to produce electricity. Essentially, it operates on photovoltaic (PV) systems, which directly transform sunlight into electrical energy through solar panels. These panels, also known as solar modules, are made up of multiple solar cells composed of semiconductor materials such as silicon.

The journey of converting solar energy into electricity commences as sunlight makes contact with the surface of solar panels. Sunlight, comprised of photons, acts as particles carrying light energy. The interaction between photons and the surface of solar cells within panels results in the displacement of electrons from the semiconductor material, leading to the generation of an electric current. This flow of electrons generates direct current (DC) electricity within the solar cells.

To make this electricity usable for powering homes, businesses, or other electrical devices, it needs to be converted from DC to alternating current (AC) electricity. This conversion is achieved using an inverter, which is typically installed alongside the solar panels. The inverter is responsible for converting the direct current (DC) electricity that is produced by the solar panels into alternating current (AC) electricity, making it compatible with standard electrical systems to power appliances and devices.

Installing solar panels involves several key steps to ensure optimal performance and efficiency:

1. **Site Assessment:** Before installing solar panels, it's important to assess the site to determine its suitability for solar energy generation. Elements like the alignment of the roof, presence of obstructions like trees or buildings casting shade, and the prevailing weather conditions in the area can influence how effectively solar panels operate. Conducting a thorough site evaluation is essential for identifying the optimal position and alignment to ensure maximum exposure to sunlight.

2. **Design and Planning:** After evaluating the site, the subsequent stage entails crafting the solar energy system's design. This encompasses sizing and arranging the solar array, choosing suitable solar panels and inverters, and devising the electrical wiring and mounting setup. Engaging a skilled solar installer or engineer is advisable to facilitate the design and planning process, guaranteeing that the system aligns with the site's distinct needs and specifications.

3. **Permitting and Approvals:** Before installing solar panels, it's necessary to obtain any required permits and approvals from local building authorities or utility companies. This may involve

submitting plans and documentation for review and obtaining permits for electrical work and structural modifications if needed.

4. **Installation:** Once all permits and approvals have been obtained, the solar panels can be installed on the roof or ground-mounted in the designated location. The installation process typically involves securing mounting brackets or racks to the roof or ground, attaching the solar panels to the mounting system, and wiring the panels together and to the inverter. Proper installation is critical for ensuring the safety, durability, and performance of the solar energy system.

5. **Connection to the Grid:** Solar energy systems are often linked to the electrical grid, enabling surplus electricity produced by the solar panels to be exported to the grid in exchange for credit or compensation. This process, known as net metering, involves installing a bi-directional meter that measures both electricity consumption from the grid and electricity generation from the solar panels. Once the solar energy system is connected to the grid and operational, it can start generating clean and renewable electricity for use on-site or export to the grid.

Battery Storage and Power Management

Battery storage and power management are integral components of off-grid or grid-tied solar energy systems, giving consumers the ability to store extra electricity generated by solar panels so that they can use it during times when there is less sunlight or when there is a strong demand for electricity. The direct current (DC) electricity that is generated by solar panels is stored in battery storage systems, and when it is required to power appliances, lights, and other electrical loads in the home, it is converted back into alternating current (AC) electricity. Here's an overview of battery storage and power management in solar energy systems:

Battery Storage Systems

Battery storage systems comprise rechargeable batteries designed to store surplus electricity generated by solar panels when solar irradiation is high. These batteries commonly utilize lithium-ion, lead-acid, or other chemistries for efficient and dependable energy storage. These systems vary in size, ranging from small-scale units intended for residential usage to large-scale configurations suitable for commercial or utility-scale purposes.

Types of Battery Storage Systems:

- **Off-Grid Systems:** Off-grid battery storage setups are engineered to supply power to residences or structures that lack connection to the conventional electrical grid. These systems rely entirely on solar panels and battery storage to meet electricity demand, making them ideal for remote or rural locations where grid connection is not feasible or cost-effective.
- **Grid-Tied Systems with Battery Backup:** Grid-tied battery storage systems link to both the electrical grid and solar panels, enabling the storage of surplus solar energy in batteries for utilization during grid outages or times of peak electricity demand. These systems provide backup power during emergencies and can also help reduce electricity bills by storing and using solar energy when grid electricity prices are high.

Power Management

Efficient power management plays a vital role in optimizing the effectiveness, dependability, and functionality of battery storage systems. Key aspects of power management include:

- **Charge Controller:** The Charge Controller maintains the balance of electrical flow between solar panels and batteries, safeguarding against overcharging and optimizing battery longevity and efficiency. Utilizing MPPT (Maximum Power Point Tracking) technology, these controllers are prevalent in solar energy setups to enhance energy extraction from solar panels.
- **Battery Management System (BMS):** Battery management systems monitor and control the charging and discharging of batteries, balancing cell voltages, protecting against overcharging and overdischarging, and optimizing battery performance and longevity. BMSs are essential for ensuring the safety and reliability of battery storage systems.
- **Inverter:** Inverters convert DC electricity stored in batteries into AC electricity for use in homes or buildings. In grid-tied systems with battery backup, inverters also synchronize with the grid to ensure seamless transition between grid and battery power during outages or fluctuations in solar energy production.
- **Energy Monitoring and Control:** Energy monitoring and control systems track electricity generation, consumption, and storage in real-time, providing users with valuable insights into system performance and energy usage patterns. These systems allow users to optimize energy management strategies and adjust settings as needed to maximize efficiency and savings.

Maximizing Efficiency in Solar Systems

Efficiency is paramount in off-grid living, where energy independence relies on harnessing the sun's power effectively. To fully optimize your solar power system, a combination of thoughtful practices, regular maintenance, and technological advancements is crucial. This section explores key strategies for maximizing efficiency in solar systems.

1. **Optimal System Sizing:** Properly sizing the solar array, battery storage, and other components of the off-grid system is essential for maximizing efficiency. Conducting a thorough assessment of energy needs, solar resources, and load requirements helps determine the appropriate system size to meet demand while minimizing excess capacity and costs.
2. **High-Efficiency Solar Panels:** Investing in high-efficiency solar panels with greater power output per square foot can maximize energy generation and reduce the required size of the solar array. Monocrystalline and polycrystalline silicon panels are commonly used in off-grid solar systems due to their reliability and efficiency.
3. **Tilt and Orientation:** Properly orienting and tilting solar panels to maximize sunlight exposure throughout the day optimizes energy harvest and efficiency. Panels should be oriented towards the sun's path and tilted at an angle equal to the local latitude for maximum solar gain, with adjustments made seasonally for optimal performance.
4. **MPPT Charge Controllers:** Maximum Power Point Tracking (MPPT) charge controllers are more effective than typical pulse width modulation (PWM) controllers because they continuously

alter the voltage and current output from solar panels to fit the charging needs of the battery. MPPT controllers can improve energy harvest by up to 30%, making them ideal for off-grid solar systems.

5. **Efficient Battery Storage:** Selecting high-quality and high-capacity batteries with efficient charge-discharge cycles and low self-discharge rates is essential for maximizing energy storage and system reliability. Lithium-ion batteries surpass traditional lead-acid batteries in terms of energy density, longevity, and overall performance.

6. **Proper Battery Management:** Optimizing battery performance can be accomplished by the implementation of a battery management system (BMS), which is designed to track and regulate the charging and discharging of batteries, prevent overcharging or overdischarging, and extend battery lifespan. BMSs also ensure safe and efficient operation of the off-grid solar system.

7. **Energy Monitoring and Optimization:** Installing energy monitoring systems to track energy generation, consumption, and storage in real-time allows users to identify inefficiencies, adjust system settings, and optimize energy management strategies for maximum efficiency and savings.

8. **Energy-Efficient Appliances:** Using energy-efficient appliances, LED lighting, and low-power electronics reduces energy demand and increases the overall efficiency of the off-grid solar system. Selecting appliances with ENERGY STAR ratings and low standby power consumption helps minimize energy usage and extend battery life.

9. **Load Management:** Implementing load management strategies such as scheduling high-energy tasks during peak solar production hours and prioritizing essential loads helps balance energy supply and demand, optimize battery usage, and maximize system efficiency.

10. **Regular Maintenance:** Performing routine maintenance tasks such as cleaning solar panels, inspecting wiring and connections, and testing battery health ensures optimal performance and efficiency of the off-grid solar system. Regular maintenance helps identify and address issues promptly, preventing downtime and maximizing system lifespan.

Chapter 13
Alternative Energy Sources

In no-grid situations, alternative energy sources beyond solar power can provide reliable and sustainable electricity. Relying solely on solar power may not be sufficient, especially in regions with varying weather patterns. This chapter explores alternative energy sources, highlighting the significance of diversification for enhanced resilience and self-sufficiency. From wind, hydro, and biomass energy to DIY energy projects and innovations, as well as the delicate balance of multiple energy sources, this chapter unveils a spectrum of possibilities for a robust off-grid energy portfolio.

Wind, Hydro, and Biomass Energy

Wind, hydro, and biomass energy are three alternative sources of renewable energy that can play a vital role in providing electricity in off-grid or remote areas where traditional grid access is limited. Each of these energy sources harnesses natural elements to generate power in environmentally friendly ways.

Wind Energy

With the use of wind turbines, wind energy is able to effectively harness the power of the wind and produce electricity. These turbines, often seen in open fields or wind farms, consist of tall towers with blades attached to a rotor. When wind gusts, it propels the rotor into motion, transforming wind's kinetic energy into mechanical energy. Subsequently, a generator ensconced within the turbine converts this mechanical energy into electricity.

Wind energy is abundant and widely available, particularly in coastal areas, plains, and mountain passes where wind speeds are consistently high. Wind turbines have the flexibility to be installed either individually or in groups, where larger turbines possess the capacity to generate substantial electricity. The performance of a wind turbine is contingent upon variables such as wind velocity, blade dimensions, and turbine elevation.

Despite its benefits, wind energy also faces challenges such as intermittency and variability in wind patterns. In response to these obstacles, battery storage and pumped hydro storage are two examples of energy storage methods that are regularly used in wind energy systems. These systems store surplus energy produced during peak wind periods and release it during times of high demand or low wind speeds.

Hydroelectric Energy

Hydroelectric power utilizes the kinetic energy of flowing water, found in rivers or streams, to produce electricity via hydroelectric dams or run-of-river setups. In a hydroelectric dam, water is stored in a reservoir behind the dam and released through turbines to generate electricity. In run-of-river systems, the natural flow of the river is used to turn turbines without the need for a reservoir.

Hydroelectric energy is a reliable and consistent source of renewable energy, as water flow can be controlled to match electricity demand. Large-scale hydroelectric dams can produce significant amounts of electricity and provide additional benefits such as flood control, irrigation, and recreation. Nevertheless, the establishment of sizable dams may entail environmental and societal repercussions, such as habitat disturbance and community displacement.

Micro-hydro systems offer a smaller-scale alternative to large dams, using natural waterways to generate electricity for off-grid or remote locations. These systems are often less disruptive to the environment and can be installed in a variety of settings, including small rivers, streams, or irrigation canals.

Biomass Energy

Biomass energy harnesses organic substances like wood, agricultural leftovers, or biofuels to generate heat, electricity, or biofuels via combustion or conversion methods. Biomass can either be directly burned for heating homes or powering industries, or it can undergo transformation into biogas, biodiesel, or ethanol using techniques like anaerobic digestion or pyrolysis.

Biomass energy is a versatile and readily available source of renewable energy, as biomass materials are abundant and can be sourced locally. Biomass energy systems can provide reliable heat and electricity in off-grid settings, particularly in rural areas where agricultural residues or forestry waste are abundant.

However, biomass energy also raises concerns about air quality, deforestation, and competition with food crops for land and resources. Sustainable biomass practices, such as using waste materials or cultivating dedicated energy crops on marginal lands, can help mitigate these concerns and ensure the long-term viability of biomass energy as a renewable energy source.

DIY Energy Projects and Innovations

DIY energy projects and innovations empower individuals to create their own renewable energy solutions, fostering energy independence and sustainability. Here are several DIY energy projects that enthusiasts can undertake to harness clean and renewable energy:

1. **Solar-Powered Charger:** Build a solar-powered charger for small electronic devices like smartphones, tablets, or rechargeable batteries. This project typically involves connecting a small solar panel to a charge controller and a battery bank, allowing users to charge their devices using sunlight.
2. **Solar Water Heater:** Construct a DIY solar water heater to provide hot water for household use. This project typically involves building a simple solar collector using black-painted pipes or cans to absorb sunlight and heat water. The heated water can then be stored in an insulated tank for later use.
3. **Wind Turbine Generator:** Build a small-scale wind turbine generator to generate electricity from wind power. This project typically involves constructing a rotor with blades, a generator, and a tower to mount the turbine. DIY wind turbines can be used to power small appliances, lights, or charging stations.
4. **Bicycle Generator:** Convert a bicycle into a pedal-powered generator to produce electricity while exercising. This project involves attaching a generator or alternator to the bicycle's wheel and connecting it to a battery bank or electrical devices. Pedal-powered generators can be used to charge batteries, power lights, or operate small appliances.

5. **Hydroponic Solar Pump:** Create a DIY hydroponic solar pump to circulate water in a hydroponic gardening system using solar power. This project typically involves building a small solar-powered water pump using a solar panel, a submersible pump, and tubing to deliver water to the hydroponic system.

6. **Solar Oven:** Build a solar oven to cook food using sunlight as a heat source. This project typically involves constructing a box-like structure with reflective surfaces to focus sunlight onto a cooking chamber. Solar ovens can be used to bake, roast, or steam food without the need for conventional fuel sources.

7. **DIY Solar Lighting:** Install DIY solar lighting systems to illuminate outdoor spaces such as gardens, pathways, or patios. This project typically involves using solar-powered LED lights connected to a solar panel and rechargeable batteries. DIY solar lighting systems are easy to install and require no wiring or electricity.

8. **Solar Air Heater:** Construct a DIY solar air heater to supplement home heating using solar energy. This project typically involves building a solar collector with a black absorber surface and a fan to circulate heated air into the living space. Solar air heaters can help reduce heating costs and reliance on fossil fuels during the winter months.

9. **DIY Solar Still:** Create a DIY solar still to purify water using sunlight. This project typically involves building a simple distillation apparatus using a transparent cover, a collection container, and a heat-absorbing surface. Solar stills can be used to produce clean drinking water from contaminated or brackish sources.

10. **Solar-Powered Aquaponics System:** Set up a DIY solar-powered aquaponics system to grow vegetables and raise fish using renewable energy. This project combines hydroponic gardening with aquaculture, using solar power to circulate water and provide lighting for plant growth.

These DIY energy projects and innovations demonstrate the potential for individuals to harness clean and renewable energy through creativity, innovation, and hands-on experimentation. Whether building solar chargers, wind turbines, or solar ovens, DIY enthusiasts can contribute to a more sustainable future by exploring and implementing their own renewable energy solutions.

Balancing Multiple Energy Sources

Integrating multiple energy sources presents both challenges and opportunities for off-grid living, requiring careful planning and management to achieve optimal results. Here are several strategies for effectively balancing multiple energy sources:

1. **Complementary Resource Matching:** Identify the complementary characteristics of different energy sources and match them to the energy demand profile. For example, solar energy is abundant during the day but decreases at night, while wind energy may be more consistent but varies in intensity. By combining these sources, users can leverage their strengths and compensate for their limitations to maintain a stable energy supply.

2. **Hybrid Systems Integration:** Integrate multiple energy sources into hybrid systems that combine the strengths of each source to maximize energy production and reliability. Hybrid systems often include solar panels, wind turbines, and battery storage, with backup generators or grid connections for additional reliability. Advanced control algorithms and energy management

systems can optimize the operation of hybrid systems, dynamically adjusting energy generation and storage based on changing conditions.

3. **Diversification of Energy Sources:** Diversify energy sources to reduce reliance on a single energy source and increase resilience to fluctuations or disruptions. Incorporate a mix of renewable energy sources such as solar, wind, hydro, and biomass, as well as energy storage technologies like batteries or pumped hydro storage. By diversifying energy sources, users can spread risk and ensure continuous energy supply under varying conditions.

4. **Energy Storage and Management:** Utilize energy storage options like batteries, pumped hydro storage, or thermal storage systems to store surplus energy generated during peak production times, enabling its use during off-peak periods or times of heightened demand. Energy management systems optimize the charging and discharging of storage systems based on energy availability, demand patterns, and user preferences, ensuring efficient use of stored energy and maximizing system performance.

5. **Load Management and Demand Response:** Manage energy demand through load shifting, demand response, and energy efficiency measures to align energy consumption with available energy supply. Schedule high-energy tasks such as heating, cooling, or charging during periods of abundant energy production and reduce energy consumption during periods of low production or high demand. Smart appliances, programmable thermostats, and energy monitoring systems can help optimize energy use and reduce waste.

6. **Flexibility and Redundancy:** Design energy systems with flexibility and redundancy to accommodate fluctuations in energy production and demand. Incorporate backup energy sources such as generators, grid connections, or redundant components to provide additional reliability and resilience. Flexible energy systems can adapt to changing conditions and maintain continuous operation under various scenarios, ensuring uninterrupted energy supply.

PART VII
COMMUNICATION AND COMMUNITY

Chapter 14
Off-Grid Communication

Off-grid communication is the lifeline that ensures connectivity, safety, and community resilience in remote living environments. As the world becomes increasingly interconnected, the ability to communicate effectively in remote areas becomes not just a convenience but a necessity for a sustainable and connected off-grid lifestyle. This chapter explores the significance of off-grid communication, delving into the establishment of reliable communication systems, developing emergency communication protocols, and harnessing modern communication technologies for resilient connectivity.

Establishing Reliable Communication Systems

In remote off-grid locations, establishing reliable communication systems is essential for staying connected with the outside world and maintaining contact within the community. stablishing reliable communication systems involves a multifaceted approach that combines traditional and modern technologies.

1. **Two-Way Radios and Walkie-Talkies**

Two-way radios and walkie-talkies remain essential tools for off-grid communication due to their simplicity, portability, and direct communication capabilities. These devices operate on specific radio frequencies and enable instant communication within a limited range, making them ideal for small communities or groups.

When establishing reliable communication with two-way radios:

- **Choose the Right Model:** Select radios with sufficient range for your needs, considering the terrain and potential obstacles. Models with features like weather channels and emergency alerts add versatility.
- **Coordinate Frequencies:** Ensure that all members of the community or group are using the same radio frequency to facilitate effective communication. Establish specific channels for different purposes, such as emergency communication or routine check-ins.
- **Consider Licensing:** Check local regulations regarding radio frequency usage. In some cases, obtaining the appropriate licenses may be necessary, especially for long-range or high-powered radios.

2. **Satellite Communication Devices**

Satellite communication devices provide a broader reach, offering connectivity even in the most remote off-grid locations where traditional cellular networks are unavailable. Key devices in this category include satellite phones, handheld satellite messengers, and personal locator beacons (PLBs). These devices use satellites in orbit to transmit signals, ensuring communication in areas without terrestrial coverage.

Consider the following when integrating satellite communication devices:

- **Device Selection:** Choose devices based on your specific needs. Satellite phones provide voice communication, messengers offer text messaging and location sharing, while PLBs are designed for emergency distress signaling.
- **Subscription Plans:** Subscribe to appropriate service plans based on your expected usage. Plans may vary in terms of coverage, data allowances, and cost, so select one that aligns with your communication requirements.
- **Emergency Preparedness:** Keep satellite communication devices fully charged and incorporate them into emergency communication protocols. Ensure that all community members are familiar with their usage and know how to summon help in case of emergencies.

3. **HAM Radio and Amateur Radio**

HAM radio, or amateur radio, provides long-range communication capabilities and is well-suited for off-grid environments. To operate HAM radios legally, individuals must obtain the necessary licenses. HAM radio networks can be established within off-grid communities, enabling communication over substantial distances, sharing information, and accessing emergency services when needed.

Key considerations for incorporating HAM radio into off-grid communication:

- **Licensing Requirements:** Obtain the appropriate licenses for operating HAM radios. Different license classes may offer varying privileges, so choose the one that aligns with your communication goals.
- **Equipment Selection:** Invest in reliable HAM radio equipment suitable for off-grid use. Consider factors such as frequency bands, power output, and portability when selecting radios.
- **Training and Practice:** Ensure that community members using HAM radios are trained in their operation. Regular practice sessions can enhance proficiency and familiarity with emergency communication protocols.

Emergency Communication Protocols

Off-grid living necessitates the development of robust emergency communication protocols to handle unexpected situations and ensure the safety of individuals and communities. Here are key elements to include in emergency communication protocols:

1. **Emergency Contacts:** Put together a list of people to contact in case of an emergency, comprising local authorities and medical facilities, fire departments, and relevant organizations. Ensure all residents have access to these contact numbers and know when and how to use them in emergencies.
2. **Designated Communication Channels:** Establish designated communication channels or frequencies for emergency communication. This may include specific radio channels, satellite phone numbers, or emergency communication apps. Clearly communicate these channels to all residents and ensure everyone knows how to access them.

3. **Emergency Alert System:** Implement an emergency alert system to quickly disseminate critical information to residents during emergencies. This may involve using sirens, loudspeakers, or automated messaging systems to notify residents of imminent threats or evacuation orders.
4. **Emergency Communication Plan:** Develop a comprehensive emergency communication plan outlining procedures for initiating, coordinating, and terminating emergency communication efforts. Specify roles and responsibilities for designated communication coordinators or emergency response teams.
5. **Check-In Procedures:** Establish check-in procedures for residents to confirm their safety and whereabouts during emergencies. This may involve using predetermined signals, codes, or check-in locations to account for all residents and identify individuals in need of assistance.
6. **Evacuation Routes and Assembly Points:** Identify evacuation routes and assembly points where residents should gather during emergencies. Ensure all residents are familiar with these locations and know how to reach them safely.
7. **Information Dissemination:** Define protocols for disseminating critical information to residents, including evacuation orders, shelter locations, emergency services availability, and safety instructions. Utilize multiple communication channels to reach residents, including verbal announcements, written notices, and digital alerts.
8. **Backup Communication Systems:** Establish backup communication systems in case primary communication channels are unavailable during emergencies. This may include alternative radio frequencies, satellite phones, or manual signaling methods such as whistles or flags.
9. **Training and Drills:** Provide training and conduct regular drills to familiarize residents with emergency communication protocols and procedures. Practice scenarios such as evacuations, medical emergencies, and natural disasters to ensure residents are prepared to respond effectively in real-life situations.
10. **Review and Update:** Regularly review and update emergency communication protocols based on lessons learned from drills, feedback from residents, and changes in community needs or infrastructure. Ensure protocols remain relevant and effective in addressing emerging threats or challenges.

Modern Communication Technologies

Advancements in modern communication technologies offer off-grid individuals and communities unprecedented opportunities to stay connected, informed, and empowered. By leveraging these technologies effectively, off-grid residents can enhance their quality of life and resilience in remote environments.

1. **Satellite Internet**

Satellite internet services provide off-grid locations with access to high-speed broadband internet, overcoming the limitations of traditional terrestrial infrastructure. Utilizing geostationary satellites positioned in orbit around the Earth, satellite internet providers beam internet signals directly to satellite dishes installed at subscribers' locations. This technology enables off-grid residents to access online resources, communicate via email, participate in video calls, and stream media content, regardless of their geographic location.

Satellite internet offers significant advantages for off-grid living, providing broadband access to remote areas where terrestrial internet options are unavailable or impractical. It enables off-grid individuals and communities to stay connected online, access educational resources, conduct business activities, and engage with the global community. However, satellite internet services may have limitations such as latency issues and data caps, which can affect real-time communication and streaming activities.

2. Voice over IP (VoIP) Services

Voice over IP (VoIP) services utilize internet connectivity to enable voice and video calls, text messaging, and multimedia communication over IP networks. Popular VoIP applications and platforms include Skype, WhatsApp, Zoom, and Google Meet, among others. Off-grid residents can leverage VoIP services to communicate with family, friends, and colleagues around the world, utilizing internet-based communication tools for both everyday communication and emergency coordination.

VoIP services offer several advantages for off-grid communication, including cost-effectiveness, versatility, and scalability. By utilizing VoIP applications on smartphones, tablets, or computers, off-grid individuals can make voice and video calls, send text messages, and share multimedia content over the internet, even in remote environments with limited cellular coverage. Additionally, VoIP services often provide features such as call forwarding, voicemail, and conference calling, enhancing communication capabilities for off-grid residents.

3. Cellular Signal Boosters

Cellular signal boosters amplify weak cell signals, extending coverage and improving connectivity in off-grid areas with marginal cellular reception. These devices consist of an external antenna that captures weak cellular signals from nearby cell towers, a signal amplifier that increases the power of the signal, as well as an internal antenna that rebroadcasts the signal that has been amplified within the building or vehicle. Cellular signal boosters enhance cellular connectivity for voice calls, text messages, and mobile data services, enabling off-grid residents to stay connected via their cellular devices.

Cellular signal boosters offer significant benefits for off-grid communication, particularly in areas with marginal cellular coverage. By amplifying weak cell signals, these devices improve voice call quality, reduce dropped calls, and enhance data speeds, providing off-grid residents with a more reliable and consistent cellular connection. Additionally, cellular signal boosters are easy to install and compatible with all major cellular carriers, making them a versatile solution for improving cellular connectivity in remote environments.

4. Mesh Networks

Mesh networks utilize peer-to-peer communication to create decentralized connectivity, enabling off-grid communities to establish local communication networks without reliance on centralized infrastructure. In a mesh network, each node serves as a relay point, forwarding data packets to neighboring nodes until they reach their intended destination. This distributed architecture ensures robust and resilient communication, even in the absence of traditional internet access.

Mesh networks offer several advantages for off-grid communication, including scalability, flexibility, and self-healing capabilities. By deploying mesh network nodes within the community, off-grid residents can create a resilient communication infrastructure that adapts to changing conditions and extends coverage as needed. Mesh networks can support various communication applications, including voice calls, text messaging, file sharing, and internet access, enabling off-grid communities to stay connected and informed in remote environments.

Modern communication technologies offer off-grid individuals and communities unprecedented opportunities to stay connected, informed, and empowered in remote environments. From satellite internet to VoIP services, cellular signal boosters, and mesh networks, these technologies enable off-grid residents to overcome the limitations of traditional infrastructure and establish reliable communication systems that enhance resilience and quality of life. By leveraging modern communication technologies effectively, off-grid communities can stay connected with the outside world, access essential services, and collaborate with peers, contributing to a sustainable and thriving off-grid lifestyle.

Chapter 15
Building a Supportive Community

In the realm of off-grid living, community is not just a social construct but a lifeline, offering shared resources, collective knowledge, and a sense of belonging. The significance of building a supportive community in off-grid environments goes beyond mere cooperation; it becomes a foundation for resilience, self-sufficiency, and mutual aid. This chapter explores the importance of community building in off-grid living, delving into strategies such as creating alliances and sharing resources, organizing community events and workshops, and fostering collaborative projects for sustainable community development.

Creating Alliances and Sharing Resources

Creating alliances and sharing resources are fundamental principles in fostering resilience and sustainability in off-grid or remote communities. By forming alliances and pooling resources, communities can address common challenges, leverage collective strengths, and build a supportive network that benefits all members.

Skill and Knowledge Exchange

One of the fundamental pillars of creating alliances in off-grid communities is the exchange of skills and knowledge. Every community member brings a unique set of experiences, expertise, and capabilities to the table. By fostering a culture of skill and knowledge exchange, off-grid communities can tap into this collective wisdom to address a myriad of challenges.

Communities can organize workshops, training sessions, and skill-sharing events where individuals with specific expertise, whether in renewable energy systems, permaculture, water harvesting, or construction techniques, share their knowledge with others. This exchange not only facilitates the acquisition of practical skills but also nurtures a collaborative learning environment where community members become both teachers and students.

For example, a member with expertise in solar energy systems might conduct a workshop on designing and maintaining off-grid solar setups. Another member with permaculture knowledge could share insights on sustainable farming practices. Through these exchanges, the community collectively builds a diverse skill set, reducing reliance on external expertise and promoting self-sufficiency.

Resource Sharing and Bartering

Resource sharing and bartering within off-grid communities are mechanisms that contribute to the equitable distribution of essential items, tools, and materials. In an environment where individual self-sufficiency is crucial, the ability to share resources becomes a powerful tool for overcoming challenges and optimizing the use of available assets.

Communities can establish systems for sharing tools, equipment, seeds, and even surplus food or energy. For instance, if one household has excess produce from their garden, they can share it with neighbors who may have surplus firewood or tools. This not only reduces individual expenses but also fosters a sense of mutual support and interdependence.

Bartering, the exchange of goods or services without the use of money, is another practice that strengthens resource-sharing dynamics. Community members can trade items or skills, creating a system where everyone benefits. This not only maximizes the utility of available resources but also builds social bonds and a sense of reciprocity within the community.

For example, a community member skilled in carpentry might exchange their services for fresh produce from a neighbor's garden. This not only fulfills individual needs but also cultivates a sense of cooperation, making the community more resilient to fluctuations in resource availability.

Cooperative Purchasing and Bulk Buying

Cooperative purchasing and bulk buying are strategies that off-grid communities can employ to reduce costs and access essential supplies more affordably. By leveraging the collective purchasing power of the community, individuals can negotiate discounts, obtain better deals, and enhance overall economic efficiency.

Communities can organize joint efforts to procure items such as solar panels, batteries, building materials, or other essentials in bulk. This not only reduces the financial burden on individual households but also ensures that everyone has access to necessary items at a lower cost.

For instance, if multiple households within the community require solar panels for their energy systems, they can collectively purchase these items in bulk, taking advantage of quantity discounts. This cooperative approach makes sustainable technologies more accessible and cost-effective, promoting the widespread adoption of environmentally friendly practices.

Community Banking and Financial Cooperatives

Establishing community banking and financial cooperatives is another way to create alliances and share resources within off-grid communities. These financial structures enable members to pool their resources, create a communal fund, and provide financial support to individuals or projects within the community.

Community banking systems can include rotating savings and credit associations (ROSCAs), where members contribute a fixed amount regularly, and the pooled funds are rotated among members. This approach allows individuals to access funds for specific needs while promoting financial discipline and community trust.

Financial cooperatives within off-grid communities can also extend their reach to provide low-interest loans for community development projects. Whether it's investing in renewable energy infrastructure, building communal spaces, or supporting local businesses, these financial structures empower the community to leverage their collective financial resources for the greater good.

Organizing Community Events and Workshops

Community events and workshops play a vital role in fostering connection, shared learning, and a sense of belonging. These gatherings provide opportunities for individuals to come together, exchange ideas, and strengthen the bonds that form the foundation of a supportive off-grid community. Here's a step-by-step guide on how to plan and execute these gatherings effectively:

1. **Identify Community Interests:** Start by gauging the interests and needs of community members. Conduct surveys or hold informal discussions to understand what topics or activities people are interested in exploring together.
2. **Plan Engaging Events:** Based on community interests, plan a variety of events and workshops that cater to different preferences and skill levels. This could include gardening workshops, cooking classes, sustainable living seminars, arts and crafts sessions, or outdoor adventure outings.
3. **Set Clear Goals:** Define clear goals and objectives for each event or workshop. Determine what you hope to achieve, whether it's educating participants, building skills, fostering social connections, or promoting community resilience.
4. **Create a Schedule:** Develop a schedule or calendar of events, taking into account factors such as seasonality, weather conditions, and community availability. Spread out events throughout the year to maintain engagement and accommodate diverse interests.
5. **Secure Resources:** Gather the necessary resources and materials for each event or workshop. This may include equipment, supplies, venue arrangements, and any required permits or permissions.
6. **Promote the Events:** Spread the word about upcoming events and workshops through various channels such as community newsletters, social media, posters, word of mouth, and local bulletin boards. Encourage community members to invite friends and neighbors to participate.
7. **Facilitate Participation:** Create a welcoming and inclusive environment where everyone feels comfortable participating. Encourage active involvement and interaction among participants, and provide opportunities for hands-on learning and skill-building.
8. **Invite Guest Speakers or Experts:** Enhance the quality of your events by inviting guest speakers or experts to share their knowledge and expertise on relevant topics. This could include local farmers, environmentalists, artisans, or professionals in various fields.
9. **Collect Feedback:** Collect input from attendees following every event or workshop to evaluate its efficacy and pinpoint opportunities for enhancement. Utilize this feedback to enhance forthcoming events, ensuring they consistently cater to the community's needs and preferences.
10. **Celebrate Successes:** Celebrate the successes of your community events and workshops by acknowledging the contributions of participants, volunteers, and organizers. Recognize achievements, share stories of inspiration, and foster a sense of pride and accomplishment within the community.

Collaborative Projects and Community Development

Undertaking collaborative projects and engaging in community development initiatives are instrumental in building a resilient off-grid community. By working together on shared goals, community members can address challenges, enhance infrastructure, and create a thriving environment.

Community Gardens and Agriculture

Establishing community gardens and collectively managing agricultural projects is a cornerstone of community development in off-grid settings. These projects not only promote food security and self-sufficiency but also foster a sense of connection to the land and each other.

Community gardens provide a shared space where members can come together to grow fruits, vegetables, and herbs. By pooling resources such as seeds, tools, and labor, community members can create thriving gardens that produce nutritious food for all. These projects encourage collaboration, knowledge-sharing, and hands-on learning about sustainable farming practices.

Moreover, community gardens serve as gathering spaces where members can socialize, exchange ideas, and build relationships. They contribute to the overall well-being of the community by promoting physical activity, mental health, and a sense of belonging. Additionally, surplus produce can be shared or traded within the community, further strengthening social ties and food resilience.

Renewable Energy Initiatives

Collaborative projects focused on renewable energy initiatives play a crucial role in enhancing the energy independence and sustainability of off-grid communities. By investing in shared renewable energy infrastructure, such as solar arrays, wind turbines, or micro-hydro systems, community members can reduce reliance on fossil fuels and traditional grids.

These initiatives often involve collective financing, planning, and installation of renewable energy systems that serve multiple households or community facilities. By pooling resources and expertise, communities can overcome financial barriers and technical challenges associated with renewable energy adoption.

Renewable energy projects not only guarantee the availability of power that is both clean and dependable, but they also instill a sense of ownership and empowerment in the inhabitants of the community. They demonstrate the feasibility and benefits of renewable energy technologies, inspiring broader adoption and advocacy for sustainable energy practices. Moreover, surplus energy generated can be shared or sold within the community, generating additional income and promoting economic resilience.

Infrastructure Development and Maintenance

Engaging in collaborative projects focused on infrastructure development and maintenance is essential for the long-term viability and resilience of off-grid communities. These projects encompass a wide range of initiatives, including road construction, water system upgrades, waste management solutions, and communal space development.

Infrastructure projects require collective planning, coordination, and investment from community members. They often involve organizing work parties or volunteer efforts to address specific needs and challenges. By working together on these projects, communities can overcome resource limitations and achieve common goals more efficiently.

Infrastructure development and maintenance projects contribute to the overall well-being and quality of life in off-grid communities. They improve access to essential services, enhance safety and security, and create shared spaces for social interaction and recreation. Additionally, these projects strengthen community bonds and solidarity, as members come together to care for and improve their shared environment.

Educational Programs and Skill-Sharing Workshops

Organizing educational programs and skill-sharing workshops is another effective way to foster community development and resilience in off-grid settings. These initiatives provide opportunities for members to learn new skills, share knowledge, and empower themselves to meet the challenges of off-grid living.

Educational programs may cover a wide range of topics, including sustainable agriculture, renewable energy technologies, natural building techniques, and emergency preparedness. By bringing in experts or experienced practitioners, communities can access valuable resources and insights to inform their sustainable living practices.

Skill-sharing workshops allow community members to share their expertise and experiences with others. These hands-on learning opportunities create a culture of mutual support and collaboration, where individuals can learn from each other and build collective resilience. Moreover, workshops promote creativity, innovation, and self-reliance, as community members explore new ways to solve problems and address shared challenges.

Collaborative projects and community development initiatives are essential for building resilience and sustainability in off-grid communities. By working together towards shared goals, community members can leverage their collective resources, skills, and creativity to address challenges, enhance infrastructure, and foster a sense of belonging.

PART VIII
HEALTH AND WELL-BEING

Chapter 16
Health Care and Emergency Preparedness

Prioritizing healthcare and emergency preparedness is essential for those embracing off-grid living, aiming for self-sufficiency and resilience. Given the remote nature of off-grid locations and potential challenges in accessing medical facilities, thorough preparation for health emergencies is critical. This chapter explores creating a comprehensive first aid kit, managing medical crises off-grid, and addressing the crucial mental health aspects amidst isolation in off-grid environments.

Building a Comprehensive First Aid Kit

A robust first aid kit is an off-grid essential, serving as the primary line of defense against injuries and illnesses where immediate access to professional medical care may be limited. Here are the essential components to include in a comprehensive first aid kit:

- **Sterile Dressings and Bandages:** These are essential for covering wounds, cuts, and abrasions to prevent infection and promote healing. Include a variety of sizes, shapes, and types such as adhesive bandages, gauze pads, and adhesive tape.
- **Antiseptic Wipes and Cleansers:** Antiseptic wipes or solutions help clean wounds and prevent infection. Include alcohol wipes, antiseptic wipes, or antiseptic solutions like hydrogen peroxide or iodine.
- **Tweezers and Scissors:** Tweezers are useful for removing splinters, ticks, or debris from wounds, while scissors are handy for cutting bandages, clothing, or medical tape.
- **Disposable Gloves:** These gloves are designed to shield both the responder and the patient from contamination and infection during first aid interventions. Make sure to provide a range of sizes to suit various users.
- **Pain Relievers:** Utilize over-the-counter pain relievers like ibuprofen or acetaminophen to alleviate minor aches, pains, and fevers. Keep these medications stored in a waterproof container for protection.
- **Antihistamines:** Antihistamines like diphenhydramine (Benadryl) can be used to treat allergic reactions, insect bites, or stings. Include both oral tablets and topical creams or ointments for allergic reactions.
- **Antibiotic Ointment:** Antibiotic ointment like bacitracin or triple antibiotic ointment helps prevent infection in minor wounds and cuts. Apply a small amount to clean wounds before bandaging.
- **CPR Mask:** A CPR mask with a one-way valve provides a barrier between the responder and the patient during CPR, reducing the risk of cross-contamination and infection transmission.

- **Emergency Blanket:** Also referred to as a space blanket, this essential item aids in preserving body heat and averting hypothermia during emergencies. These lightweight, compact blankets are essential for treating shock or exposure to extreme temperatures.
- **Medical Tape and Bandages:** Medical tape secures dressings and bandages in place, while elastic bandages provide support and compression for sprains, strains, or injuries.
- **Eye Wash Solution:** Eye wash solution or saline solution helps rinse foreign particles or contaminants from the eyes in case of chemical exposure or eye injuries.
- **Splinting Materials:** Include splinting materials such as wooden or aluminum splints, SAM splints, or rolled-up newspapers for stabilizing fractures or injuries to limbs.
- **Emergency Contact Information:** Include a list of emergency contact numbers, medical history, allergies, and medications for each member of the household. Store this information in a waterproof container or ziplock bag.
- **First Aid Manual:** A detailed guidebook on first aid offers comprehensive instructions for evaluating and managing a wide range of injuries and medical crises. Include a manual with clear instructions and illustrations for easy reference.

When assembling a first aid kit for off-grid living, consider the specific needs and risks of the community, such as outdoor activities, remote location, and limited access to medical facilities. Regularly check and update the contents of the first aid kit to ensure it remains well-stocked and up-to-date. Additionally, provide training and instruction on how to use the first aid kit and perform basic first aid procedures to all members of the community. By building a comprehensive first aid kit and empowering residents with the knowledge and skills to use it effectively, off-grid communities can enhance their resilience and ability to respond to medical emergencies.

Handling Medical Emergencies Off-Grid

In off-grid environments, access to traditional medical facilities may be limited, making it essential to be prepared to handle medical emergencies with resourcefulness and skill. Here are some common medical emergencies and how to handle them in off-grid settings:

Cuts and Wounds

1. Clean the wound with clean water and antiseptic solution, if available.
2. Apply direct pressure using a sterile dressing or clean cloth to stop bleeding.
3. If the wound is deep or gaping, use butterfly bandages or sutures to close it.
4. Cover the wound using a sterile dressing and secure it in place with medical tape or bandage.
5. Be vigilant for indications of infection like redness, swelling, or discharge, and promptly consult medical assistance if needed.

Burns

1. Cool the burn with cold running water or immerse in cold water for at least 10 minutes to relieve pain and reduce swelling.

2. Cover the burn using a sterile dressing or clean cloth to protect it from infection.
3. Do not apply ice, butter, or other home remedies to the burn as they can worsen the injury.
4. Seek medical attention for severe burns, burns affecting large areas of the body, or burns with blistering or charring.

Fractures and Sprains

1. Immobilize the injured limb using splinting materials such as wooden boards, aluminum splints, or rolled-up newspapers.
2. Apply ice packs or cold compresses to lessen the swelling and relieve pain.
3. Elevate the injured limb above the heart if possible to reduce swelling.
4. Seek medical help for suspected fractures, dislocations, or severe sprains that may require professional treatment.

Heat Exhaustion and Heatstroke

1. Move the person to a cooler, shaded area and loosen tight clothing.
2. Offer cool fluids to drink and apply cool, wet cloths to the skin to lower body temperature.
3. Monitor for signs of heat exhaustion such as heavy sweating, weakness, dizziness, or nausea.
4. If symptoms worsen or include confusion, rapid pulse, or loss of consciousness, seek immediate medical attention as it may indicate heatstroke, a life-threatening condition.

Allergic Reactions

1. Remove the trigger if possible (e.g., bee stinger) and wash the affected area using soap and water.
2. Administer antihistamines such as diphenhydramine (Benadryl) to reduce allergic symptoms.
3. If the person has a history of severe allergic reactions (anaphylaxis), administer epinephrine if available and seek immediate medical help.

Choking

1. Encourage the person to cough forcefully to dislodge the obstruction.
2. Perform abdominal thrusts (Heimlich maneuver) if the person is conscious and unable to breathe.
3. If the person becomes unconscious, perform CPR starting with chest compressions and checking for objects in the airway after each cycle.

Heart Attack

1. Have the person rest in a comfortable position and reassure them.
2. If available, administer aspirin to chew and swallow to help reduce blood clotting.
3. Monitor vital signs then be prepared to perform CPR if the person becomes unresponsive.

Seizures

1. Protect the person from injury by taking out nearby objects and cushioning their head.
2. Do not restrain the person or put anything in their mouth.

3. Time the duration of the seizure and stay with the person until it ends.
4. After the seizure, help the person into a safe position and monitor their breathing and consciousness.

In any medical emergency, it's essential to stay calm, assess the situation, and provide appropriate first aid while awaiting professional medical assistance. Additionally, having a well-stocked first aid kit, basic medical supplies, and knowledge of first aid procedures can significantly improve the outcomes of medical emergencies in off-grid settings. Regular training and practice in first aid techniques are also recommended for all members of off-grid communities.

Mental Health Considerations in Isolation

Off-grid living can present unique challenges to mental health, including isolation, stress, and limited access to support services. Prioritizing mental well-being is essential for maintaining resilience and overall quality of life.

1. **Social Connection:** Isolation can lead to feelings of loneliness and social isolation, impacting mental health. Off-grid communities should prioritize opportunities for social connection and support among residents. This could include organizing regular community gatherings, group activities, or virtual communication channels to foster connections and combat feelings of loneliness.
2. **Physical Activity:** Engaging in regular physical activity has been proven to enhance mood and alleviate symptoms associated with depression and anxiety. Off-grid living offers ample opportunities for outdoor activities such as hiking, gardening, or nature walks, which can promote physical health and mental well-being. Encouraging residents to engage in regular exercise routines can help maintain positive mental health in isolation.
3. **Routine and Structure:** Creating a daily routine and structure offers stability and purpose, which can be particularly beneficial in isolated environments. Off-grid communities can create schedules for daily tasks, chores, and activities, helping residents stay organized and focused. Having a sense of predictability and control over daily life can reduce stress and anxiety associated with isolation.
4. **Mindfulness and Relaxation Techniques:** People who engage in stress management and the management of negative emotions may find that using mindfulness and methods of relaxation is beneficial. Off-grid communities can offer mindfulness meditation sessions, yoga classes, or relaxation exercises to promote mental well-being. Encouraging residents to prioritize self-care and stress-reduction strategies can enhance resilience in isolation.
5. **Access to Mental Health Support:** Access to mental health support services is crucial for addressing mental health concerns in remote areas. Off-grid communities must ensure access to resources like counseling services, support groups, or telehealth options to support mental well-being. Residents should be encouraged to seek help if they experience mental health challenges, and efforts should be made to reduce stigma surrounding mental illness.
6. **Connection with Nature:** Off-grid living often involves close proximity to nature, which can have therapeutic benefits for mental health. Spending time outdoors, enjoying natural surroundings, and connecting with the natural world can promote relaxation, reduce stress, and improve mood. Off-

grid communities should encourage residents to spend time in nature and appreciate its healing properties.

7. **Community Support:** Strong social support networks within off-grid communities are vital for promoting mental health and resilience. Residents should feel supported by their neighbors and community members, knowing they have someone to turn to in times of need. Building a culture of kindness, compassion, and mutual support can enhance mental well-being in isolation.

8. **Resilience Building:** Off-grid living inherently involves facing challenges and adapting to change, requiring resilience. Off-grid communities can promote resilience by encouraging problem-solving skills, fostering optimism, and providing opportunities for personal growth and development. Building resilience helps individuals cope with adversity and thrive in isolation.

By addressing these mental health considerations, off-grid communities can create environments that support residents' well-being, promote resilience, and enhance overall quality of life despite the challenges of isolation.

Chapter 17
Mental Health and Social Aspects

Living off-grid presents unique challenges to mental health and social well-being. The isolated nature of off-grid environments, coupled with the demands of self-sufficiency, can impact individuals psychologically. This chapter explores strategies for coping with isolation and stress, building a balanced lifestyle, and fostering community support for mental well-being in the context of off-grid living.

Coping with Isolation and Stress

Isolation, often romanticized as a retreat from the chaotic pace of modern life, can pose unique challenges to mental health in off-grid living. The absence of constant human interaction and the reliance on self-reliance can amplify feelings of loneliness and stress. However, there are various strategies individuals can employ to cope with these challenges:

1. **Stay Connected:** Despite being in a remote area, it's important to stay connected with others. Make use of technology like phones, emails, or video calls to keep in touch with family and friends. Engage in regular communication to share experiences, thoughts, and feelings, reducing feelings of loneliness and isolation.

2. **Establish a Routine:** A daily pattern may offer structure and consistency, so lowering emotions of anxiety and uncertainty. Establishing a routine can be beneficial. Plan out your day with regular activities such as waking up and going to bed at consistent times, mealtimes, and designated work or leisure hours. Having a routine can give you a sense of control over your day and help manage stress.

3. **Engage in Meaningful Activities:** Find activities that bring you joy and fulfillment. Whether it's gardening, painting, reading, or hiking, engaging in activities you enjoy can help distract your mind from stressors and promote relaxation. Take time to pursue hobbies or interests that nourish your soul and bring you happiness.

4. **Practice Self-Care:** Self-care should be your top priority in order to nourish both your physical and emotional well-being. If you want to maintain your body healthy, you should get plenty of rest, consume meals that are high in nutrients, and exercise on a regular basis. In order to alleviate stress and foster relaxation, it is beneficial to engage in relaxation practices such as yoga, meditation, or deep breathing exercises.

5. **Explore Nature:** Make the most of off-grid living by immersing yourself in the great outdoors. Nature's tranquil atmosphere can alleviate stress and enhance mental health. Take leisurely strolls, venture into the mountains for hikes, or simply relax outside and appreciate the splendor of the environment.

6. **Seek Support:** If you find yourself in need of assistance, do not be afraid to ask for it. Talk to trusted friends, family members, or community members about your feelings and concerns.

Expressing your thoughts and emotions to others can offer solace and affirmation, reassuring you that your experiences are shared by others.

7. **Practice Mindfulness:** Embrace the practice of being completely present in the current moment without passing judgment. Take time to observe your thoughts, emotions, and surroundings without trying to change them. Mindfulness practices such as meditation, mindfulness walks, or mindful breathing can help reduce stress and increase resilience.

8. **Limit News Consumption:** While it's important to stay informed, excessive exposure to news and media can increase feelings of anxiety and stress. Limit intake of news and social media, particularly if it's leading to feelings of distress. Focus on reliable sources of information and take breaks when needed to maintain mental well-being.

9. **Stay Positive:** Nurture a positive perspective through embracing gratitude and optimism. Dedicate time each day to acknowledge what you're thankful for. Seek out the positives within difficult circumstances, and reinforce your confidence in your capabilities to surmount challenges.

10. **Seek Professional Help:** Do not be afraid to seek the assistance of a professional if you are having difficulty coping with feelings of loneliness and stress. A therapist, counselor, or mental health practitioner can offer personalized assistance, direction, and coping mechanisms to meet your specific requirements.

By implementing these simple coping strategies, individuals in off-grid living can effectively manage isolation and stress, promoting mental well-being and resilience in remote environments.

Building a Balanced Lifestyle

Maintaining a balanced lifestyle is essential for overall well-being, particularly in the context of off-grid living where the lines between work and leisure may blur. Balancing productivity with relaxation, solitude with social interaction, and self-reliance with community support is key to fostering a fulfilling off-grid lifestyle. Here are some tips for building a balanced lifestyle:

1. **Physical Health:** Make it a priority to keep up a healthy lifestyle by consuming meals that are high in nutrients, ensuring that you drink plenty of water, and indulging in frequent physical activity. Take advantage of outdoor opportunities in off-grid settings for hiking, gardening, or other forms of exercise.

2. **Mental Well-being:** Make mental well-being a priority by incorporating stress-relieving practices like meditation, cmindfulness, and deep breathing exercises into your routine. Keep in touch with friends and family, reach out for support when necessary, and participate in activities that bring you relaxation and joy.

3. **Social Connections:** Foster meaningful social connections by building relationships with fellow community members, participating in group activities, and maintaining regular communication with friends and family. Cultivate a sense of belonging and community by supporting and connecting with others.

4. **Personal Fulfillment:** Pursue activities and interests that bring you fulfillment, joy, and a sense of purpose. Whether it's pursuing hobbies, learning new skills, or volunteering in the community, make time for activities that nourish your soul and enhance your quality of life.

5. **Work-Life Balance:** Strive for a healthy balance between work and leisure activities. Set boundaries around work commitments, prioritize self-care and relaxation, and make time for activities outside of work that bring you joy and fulfillment.

6. **Connection with Nature:** Embrace the natural surroundings in off-grid living by spending time outdoors and connecting with the natural world. Enjoy activities such as hiking, camping, or simply immersing yourself in the beauty of nature to promote relaxation and well-being.

7. **Mindful Consumption:** Practice mindful consumption by being intentional about how you use your time, energy, and resources. Avoid excessive consumption of material goods or information that may contribute to stress or overwhelm. Instead, focus on experiences and activities that add value to your life and promote well-being.

8. **Flexibility and Adaptability:** Maintain flexibility and adaptability in your lifestyle to navigate the challenges and opportunities that come with off-grid living. Be open to change, embrace new experiences, and approach challenges with resilience and a positive mindset.

Community Support for Mental Well-being

Community support plays a crucial role in promoting mental well-being in off-grid living. Here are some ways communities can support mental health:

1. **Open Communication:** Cultivate an atmosphere of openness and sincerity within the community regarding mental health matters. Establish a nurturing space where individuals can freely express their emotions, share experiences, and address concerns without apprehension of criticism or societal prejudice.

2. **Peer Support Groups:** Establish peer support groups or buddy systems where residents can connect with others who may be experiencing similar challenges or struggles. Peer support provides validation, empathy, and understanding, helping individuals feel less isolated and more supported in their mental health journey.

3. **Community Events and Gatherings:** Organize regular community events, gatherings, or social activities that promote connection, camaraderie, and belonging. These events provide opportunities for residents to come together, share experiences, and strengthen social bonds, which are essential for mental well-being.

4. **Education and Awareness:** Provide educational workshops, seminars, or guest speakers focusing on mental health awareness, coping mechanisms, and self-care techniques. Enhance community understanding of mental health matters, enabling individuals to identify signs of distress and access support when necessary.

5. **Access to Resources:** Provide access to mental health resources and support services within the community. This could include information about local counseling services, crisis hotlines, online support groups, or mental health resources available in nearby towns or cities.

6. **Training and Skill-Building:** Offer training sessions or skill-building workshops on topics such as conflict resolution, stress management, active listening, and peer support. Equip community members with practical skills and tools to effectively support each other's mental well-being.

7. **Crisis Intervention:** Develop protocols and procedures for responding to mental health crises within the community. Train designated individuals or teams in crisis intervention techniques, suicide prevention, and de-escalation strategies to provide immediate support and assistance when needed.

8. **Promote Inclusivity and Acceptance:** Foster a culture of inclusivity, acceptance, and diversity within the community. Create an environment where all residents feel valued, respected, and included, regardless of their identity, background, or mental health status.

9. **Collaboration with External Services:** Establish partnerships with external mental health services, organizations, or professionals to supplement community-based support. Collaborate with local mental health providers to offer counseling services, support groups, or outreach programs tailored to the needs of off-grid communities.

10. **Regular Check-Ins:** Conduct regular check-ins with community members to assess their mental health and well-being. Encourage residents to reach out for support if they are struggling and provide opportunities for confidential discussions or one-on-one support as needed.

PART IX
ADVANCED PROJECTS AND INNOVATIONS

Chapter 18
Specialized DIY Projects

Specialized DIY projects offer off-grid individuals the opportunity to customize their living spaces and infrastructure according to their unique needs and priorities. From enhancing food production through greenhouses and aquaponics systems to optimizing water and energy systems with advanced technologies, these projects empower individuals to achieve greater self-sufficiency and sustainability.

Building a Greenhouse and Aquaponics System

Building a greenhouse and aquaponics system can significantly enhance food production and sustainability in off-grid living. Here's a step-by-step guide for each system:

Building a Greenhouse

1. **Site Selection:** Choose a location for the greenhouse that receives plenty of sunlight throughout the day and is sheltered from strong winds. Ensure the area is level and free from obstructions.
2. **Design Planning:** Determine the size and shape of the greenhouse based on available space and intended use. Consider factors such as ventilation, access to water, and materials for construction.
3. **Gathering Materials:** Collect materials needed for the greenhouse, including PVC pipes or metal tubing for the frame, greenhouse plastic or polycarbonate panels for covering, and hardware such as screws and brackets.
4. **Constructing the Frame:** Assemble the frame of the greenhouse according to the planned design, using connectors or fittings to join the pipes or tubing. Secure the frame firmly to the ground using stakes or anchors.
5. **Installing Covering:** Attach the greenhouse plastic or polycarbonate panels to the frame, ensuring a tight and secure fit. Trim any excess material and seal the edges to prevent drafts and water leakage.
6. **Adding Ventilation:** Install vents or windows in the greenhouse to regulate temperature and humidity levels. Consider automatic vent openers or fans for efficient airflow control.
7. **Setting Up Benches and Shelves:** Add benches or shelves inside the greenhouse for holding plants, pots, and gardening supplies. Ensure they are sturdy and can withstand the weight of plants and equipment.
8. **Watering System:** Set up a watering system like drip irrigation or soaker hoses to ensure plants receive a steady supply of moisture. Connect the system to a water source and set up timers or controllers for automated watering.

Building an Aquaponics System

1. **Design Planning:** Determine the size and layout of the aquaponics system based on available space and desired production capacity. Decide on the type of aquaponics system, such as media beds, nutrient film technique (NFT), or raft systems.

2. **Gathering Materials:** Gather materials needed for the aquaponics system, including a fish tank or pond, grow beds or trays, plumbing fittings, pumps, and filtration components. Choose food-grade materials that are safe for aquatic life.

3. **Constructing the Fish Tank:** Set up the fish tank or pond in a suitable location within the greenhouse, ensuring it is large enough to accommodate the desired fish species and provides adequate water volume for the system.

4. **Installing Grow Beds:** Position the grow beds or trays above the fish tank, allowing water to flow from the tank into the grow beds through gravity or a pump. Fill the grow beds with inert growing media such as gravel, expanded clay pellets, or rock wool.

5. **Plumbing Setup:** Install plumbing fittings and pipes to create a closed-loop system that circulates water between the fish tank and grow beds. Incorporate a pump for water circulation alongside a filtration mechanism to eliminate solid waste and uphold optimal water quality.

6. **Adding Fish:** Introduce fish species suitable for aquaponics, such as tilapia, trout, or perch, into the fish tank. Regularly check the temperature, pH, ammonia, and nitrate levels of the water to maintain the ideal conditions for the health of the fish.

7. **Planting:** Plant vegetables, herbs, or fruits in the grow beds, allowing their roots to absorb nutrients from the water as it passes through the media. Choose fast-growing, nutrient-rich crops that thrive in aquaponic environments.

8. **Monitoring and Maintenance:** Regularly monitor water quality, fish health, and plant growth in the aquaponics system. Test water parameters, feed fish appropriately, and prune plants as needed to maintain a balanced ecosystem. Conduct regular maintenance tasks such as cleaning filters, replacing water, and checking for leaks or malfunctions in the system.

Advanced Water and Energy Systems

In off-grid living, developing advanced water and energy systems ensures efficient resource management and resilience in the face of environmental challenges. Here's a step-by-step guide for setting up advanced water and energy systems:

Advanced Water System

1. **Assess Water Needs:** Determine the water needs of the off-grid community, including drinking water, irrigation for gardens, and household use. Consider factors such as water sources, usage patterns, and seasonal variations in water availability.

2. **Rainwater Harvesting:** Incorporate a rainwater collection system to gather and reserve rainwater for domestic purposes. Arrange gutters and downspouts to channel rainwater into storage tanks or cisterns. Employ filters and screens to eliminate debris and impurities from the harvested water.

3. **Greywater Recycling:** Utilize a greywater recycling system to repurpose water from sinks, showers, and laundry machines for the purpose of irrigating plants or flushing toilets. Install a greywater filtration and treatment system to remove impurities and ensure water quality.

4. **Solar Water Pumping:** Use solar-powered water pumps to extract water from wells, springs, or storage tanks for distribution to homes and gardens. Install photovoltaic (PV) panels to generate electricity for pumping water, reducing reliance on fossil fuels.

5. **Water Treatment:** Install water treatment systems such as UV sterilizers, reverse osmosis filters, or ceramic filters to purify drinking water and remove contaminants. Ensure treated water meets safety standards for consumption and sanitation.

6. **Smart Water Management:** Implement smart water management technologies such as water meters, leak detection systems, and automated irrigation controllers. Use sensors and monitoring devices to track water usage, identify leaks, and optimize irrigation schedules for maximum efficiency.

Advanced Energy System

1. **Assess Energy Needs:** Determine the energy needs of the off-grid community, including electricity for lighting, appliances, heating, and cooling. Conduct an energy audit to assess current consumption patterns and identify areas for efficiency improvements.

2. **Solar Photovoltaic (PV) System:** Install a solar PV system to generate electricity from sunlight. Select high-efficiency solar panels, inverters, and batteries to maximize energy production and storage capacity. Position solar panels in optimal locations to capture maximum sunlight throughout the day.

3. **Wind Turbine System:** Consider installing a wind turbine system to harness wind energy for electricity generation. Choose a suitable location with consistent wind patterns and minimal obstructions for optimal turbine performance. Install a tower, blades, and generator to convert wind power into electricity.

4. **Micro-hydroelectric System:** Utilize micro-hydroelectric technology to generate electricity from flowing water, such as streams or rivers. Install a water turbine, penstock, and generator to harness the kinetic energy of moving water and produce renewable electricity.

5. **Battery Storage:** The incorporation of battery storage systems allows for the storage of extra energy supplied by solar, wind, or hydro systems, which can then be utilized during times of fluctuating output or high demand. Select deep-cycle batteries with sufficient capacity and durability to withstand frequent charging and discharging cycles.

6. **Energy Management System:** Implement an energy management system to monitor and control energy usage, storage, and distribution within the off-grid community. Use smart meters, energy monitors, and remote control devices to optimize energy efficiency and reduce wastage.

7. **Backup Generator:** Install backup generators powered by diesel, propane, or biodiesel as a supplemental energy source during extended periods of low renewable energy production or emergencies. Size the generator appropriately to meet peak demand and ensure reliable power supply.

Smart Technologies in Off-Grid Living

In an off-grid living, the integration of smart technologies introduces a new paradigm that combines sustainability, efficiency, and connectivity. Embracing these technologies empowers off-grid residents to optimize resource management, enhance their quality of life, and navigate the challenges of a self-sufficient lifestyle. Here's a step-by-step guide for implementing smart technologies in off-grid living:

1. **Assess Needs:** Evaluate the specific needs and priorities of the off-grid community, considering factors such as energy consumption, water usage, security, and communication.

2. **Select Smart Devices:** Choose smart devices and technologies that address identified needs and align with the off-grid lifestyle. Examples include smart thermostats for energy-efficient heating and cooling, smart meters for monitoring energy and water usage, and smart security cameras for remote monitoring.

3. **Energy Management System:** Install an energy management system to monitor and control energy usage within the off-grid community. Use smart meters, energy monitors, and programmable thermostats to optimize energy efficiency, reduce wastage, and track renewable energy production from solar, wind, or hydro systems.

4. **Water Management System:** Implement a water management system to monitor and control water usage and distribution. Install smart water meters, leak detection sensors, and automated irrigation controllers to conserve water, detect leaks, and optimize irrigation schedules for gardens and crops.

5. **Renewable Energy Monitoring:** Use smart monitoring devices and software to track renewable energy production from solar panels, wind turbines, or micro-hydro systems. Monitor energy output, battery storage levels, and system performance in real-time to optimize energy production and storage.

6. **Home Automation:** Integrate home automation technologies to control lighting, appliances, and other devices remotely. Use smart plugs, switches, and hubs to automate routines, schedule tasks, and adjust settings based on occupancy or energy availability.

7. **Communication Systems:** Implement smart communication systems to facilitate connectivity and collaboration within the off-grid community. Use satellite internet, mesh networks, or long-range radios for reliable communication, data exchange, and emergency response.

8. **Security and Surveillance:** Install smart security and surveillance systems to enhance safety and deter intruders in off-grid locations. Use smart cameras, motion sensors, and alarm systems with remote monitoring and notification capabilities for real-time surveillance and security alerts.

9. **Environmental Monitoring:** Deploy smart environmental monitoring devices to track weather conditions, air quality, and environmental parameters in off-grid environments. Use sensors and data loggers to collect and analyze environmental data for informed decision-making and resource management.

10. **Remote Management and Control:** Utilize smart home hubs, mobile apps, or cloud-based platforms for remote management and control of smart devices and systems. Monitor and adjust settings, receive alerts, and access data from anywhere with an internet connection, enhancing convenience and flexibility in off-grid living.

Chapter 19
Leveraging Technology

In today's rapidly evolving world, off-grid living demands innovative solutions to sustainably manage resources, enhance efficiency, and ensure a high quality of life. Leveraging cutting-edge technologies becomes imperative in this pursuit. Among the array of technologies available, drones, sensors, and automation systems stand out as indispensable tools in optimizing off-grid living environments. From resource monitoring to emergency response, and from energy management to staying updated with emerging technologies, these advancements offer transformative possibilities

Using Drones and Sensors for Efficiency

In off-grid living, drones and sensors can be valuable tools for enhancing efficiency, monitoring resources, and improving overall quality of life. Here's how these technologies can be leveraged:

1. **Resource Monitoring:** Drones equipped with sensors can be deployed to monitor natural resources such as water sources, forests, and agricultural land. These drones can collect data on water levels, soil moisture, vegetation health, and environmental conditions, providing valuable insights for resource management and conservation. For example, drones can survey remote areas to identify potential water sources or monitor crop health to optimize irrigation and fertilization practices.

2. **Infrastructure Inspection:** Drones can also be used for infrastructure inspection and maintenance in off-grid communities. They can conduct aerial surveys of buildings, roads, bridges, and other structures to identify defects, damage, or maintenance needs. This allows for early detection of issues and timely repairs, ensuring the safety and longevity of essential infrastructure in remote areas.

3. **Energy System Monitoring:** Sensors installed in renewable energy systems, such as solar panels, wind turbines, and micro-hydro systems, can provide real-time data on energy production, storage, and consumption. This allows off-grid residents to monitor system performance, identify inefficiencies, and optimize energy usage. Drones can be used to inspect solar panels and wind turbines for damage or debris buildup, ensuring maximum energy output and system reliability.

4. **Wildlife Monitoring:** Drones equipped with cameras and sensors can aid in wildlife monitoring and conservation efforts in off-grid areas. They can survey remote habitats, track animal populations, and monitor wildlife behavior without disturbing natural ecosystems. This information can inform conservation strategies, habitat management practices, and wildlife protection measures in off-grid communities.

5. **Emergency Response:** Drones can play a crucial role in emergency response and disaster management in off-grid living. They can be deployed to assess damage, conduct search and rescue missions, and deliver emergency supplies to remote areas inaccessible by traditional means. In the event of wildfires, drones that are fitted with thermal imaging cameras can also assist in the search

for missing persons and the identification of hotspots, facilitating timely and effective emergency response efforts.

6. **Data Analysis and Decision-Making:** The data collected by drones and sensors can be analyzed using advanced analytics tools to derive actionable insights and inform decision-making processes in off-grid communities. This data-driven approach enables residents to make informed choices regarding resource allocation, energy management, environmental conservation, and emergency preparedness.

7. **Community Engagement:** Drones can serve as educational tools and community engagement platforms in off-grid living. They can be used to capture aerial footage of community events, natural landscapes, and infrastructure projects, fostering a sense of connection and pride among residents. Drones can also be used for educational purposes, teaching residents about environmental conservation, renewable energy, and technology innovation in off-grid living.

Automation in Off-Grid Systems

Automation plays a significant role in off-grid systems by streamlining processes, optimizing resource usage, and improving overall efficiency. Below are several key areas where automation can be effectively implemented in off-grid systems:

1. **Energy Management:** Automated energy management systems can monitor energy production, storage, and consumption in off-grid renewable energy systems. Smart controllers, sensors, and algorithms can adjust energy usage based on demand, weather conditions, and battery capacity. This ensures efficient use of renewable energy resources and maximizes system performance.

2. **Water Management:** Automated water management systems can regulate water usage, distribution, and conservation in off-grid communities. Smart irrigation controllers, moisture sensors, and leak detection systems can optimize watering schedules, prevent water waste, and detect leaks in water pipes or storage tanks. This helps conserve water resources and ensures sustainable water supply for drinking, irrigation, and other needs.

3. **Home Automation:** Home automation technologies offer the capability to streamline tasks and operations in off-grid residences, enhancing both comfort and energy efficiency. Through the utilization of smart thermostats, lighting setups, and appliances, these systems can be tailored to automatically adapt settings in response to factors such as occupancy, time, or energy resources. This reduces energy consumption, lowers utility costs, and enhances overall quality of life in off-grid living.

4. **Security and Surveillance:** Automated security and surveillance systems can enhance safety and protection in off-grid communities. Smart cameras, motion sensors, and alarm systems can detect intruders, trigger alerts, and initiate response actions. Integration with mobile apps or cloud-based platforms allows residents to monitor security cameras, receive notifications, and remotely control security devices from anywhere with an internet connection.

5. **Environmental Monitoring:** Automated systems for environmental monitoring have the capability to monitor various environmental factors like humidity, temperature, air quality, and pollution levels even in locations without access to conventional power grids. Sensors and data loggers can collect real-time data and transmit it to a central monitoring station or cloud-based

platform for analysis. This helps residents assess environmental conditions, identify trends, and take proactive measures to mitigate risks or impacts on health and well-being.

6. **Communication Systems:** Automated communication systems can facilitate connectivity and collaboration among off-grid residents. Satellite internet, mesh networks, or long-range radios can provide reliable communication channels for voice, data, and emergency communication. Automated routing, encryption, and redundancy mechanisms ensure robust and secure communication in remote areas with limited infrastructure.

7. **Emergency Response:** Automated emergency response systems can expedite emergency alerts, notifications, and coordination in off-grid communities. Integrated alarm systems, GPS trackers, and mobile apps can trigger alerts in case of emergencies, such as natural disasters, medical incidents, or security threats. Automated response protocols and communication channels help coordinate emergency response efforts and mobilize resources effectively.

Staying Updated with Emerging Technologies

Embracing the off-grid lifestyle requires a proactive approach to stay abreast of cutting-edge technologies that can elevate sustainability, efficiency, and overall quality of life. Here are several strategies off-grid residents can employ to stay informed about emerging technologies:

1. **Research and Exploration:** Regularly conduct research and explore emerging technologies relevant to off-grid living through online resources, industry publications, and research reports. Stay informed about advancements in renewable energy, water management, automation, communication, and other key areas.

2. **Industry Events and Conferences:** Attend industry events, conferences, and trade shows focused on renewable energy, sustainability, and off-grid living. These events provide opportunities to learn about new technologies, network with experts and practitioners, and gain insights into emerging trends and developments.

3. **Online Communities and Forums:** Join online communities, forums, and social media groups dedicated to off-grid living, renewable energy, and sustainable technology. Engage with other off-grid residents, share experiences, ask questions, and exchange information about emerging technologies and best practices.

4. **Professional Associations:** Become a member of professional associations or organizations related to renewable energy, sustainability, and off-grid living. These associations often provide access to industry publications, research findings, training programs, and networking opportunities to stay updated with emerging technologies and trends.

5. **Consult Experts and Advisors:** Seek advice and guidance from experts, consultants, and advisors specializing in off-grid living and sustainable technology. These professionals can provide insights, recommendations, and assistance in evaluating and implementing emerging technologies tailored to specific off-grid needs and requirements.

6. **Technology Demonstrations and Pilot Projects:** Participate in technology demonstrations, pilot projects, or field trials of emerging technologies relevant to off-grid living. Collaborate with technology providers, research institutions, or government agencies to test and evaluate new products, systems, or solutions in real-world off-grid environments.

7. **Continuous Learning and Education:** Stay committed to continuous learning and education by attending workshops, training programs, and webinars on emerging technologies and sustainability. Keep abreast of advancements in renewable energy, water management, automation, communication, and other relevant fields to stay ahead of the curve.
8. **Collaboration and Partnerships:** Foster collaboration and partnerships with technology providers, startups, researchers, and innovators working on emerging technologies for off-grid living. Explore opportunities for joint projects, knowledge sharing, and co-development efforts to leverage expertise and resources in advancing sustainable solutions.

PART X
TRANSITION AND ADAPTATION

Chapter 20
Making the Transition

Embarking on the journey to off-grid living signifies a profound shift towards self-sufficiency, sustainability, and a deeper connection with nature. However, this transition necessitates meticulous planning and execution to ensure a smooth and successful integration into a lifestyle independent of centralized utilities. From assessing your needs to overcoming initial challenges and carefully managing your finances, every step requires thoughtful consideration and preparation.

Planning and Executing Your Move

The decision to transition to off-grid living requires careful planning and strategic execution. From selecting an appropriate location to designing your off-grid home, this section provides insights into the key steps involved in making a successful transition.

1. **Assess Your Needs:** Begin by assessing your energy, water, and food requirements to determine the resources you'll need to sustain yourself off-grid. Consider factors such as terrain, climate, and available natural resources in your chosen location.

2. **Research and Learn:** Take the time to research off-grid living, renewable energy systems, water conservation methods, and sustainable food production techniques. Learn about the challenges and benefits of living off-grid to make informed decisions.

3. **Choose a Location:** Select a suitable location for your off-grid homestead based on factors such as access to sunlight, water sources, and arable land for gardening. Consider proximity to amenities, healthcare facilities, and emergency services for safety and convenience.

4. **Plan Your Energy System:** Based on the amount of energy you require and the resources you have available, you should make a decision regarding the sort of renewable energy system that you will implement, such as solar panels, wind turbines, or micro-hydro systems. Calculate the size and capacity of your energy system to meet your daily requirements.

5. **Set Up Water Management:** Implement water management systems for harvesting rainwater, recycling greywater, and conserving water usage. Install tanks, filters, and pumps to collect and distribute water for drinking, irrigation, and household needs.

6. **Establish Food Production:** Plan and set up systems for growing your own food through gardening, permaculture, or aquaponics. Prepare garden beds, plant seeds, and cultivate crops that are suitable for your climate and soil conditions.

7. **Prepare Your Home:** Make necessary modifications to your home or build a new off-grid dwelling that is energy-efficient, environmentally friendly, and self-sustaining. Install insulation, energy-efficient appliances, and passive solar features to reduce energy consumption.

8. **Acquire Necessary Tools and Supplies:** Stock up on essential tools, equipment, and supplies for off-grid living, including hand tools, gardening implements, water filtration systems, and

emergency supplies. Invest in renewable energy systems, water tanks, and storage solutions to ensure self-sufficiency.

9. **Practice Skills and Adaptation:** Learn basic survival skills such as fire-making, foraging, and first aid to prepare for emergencies and unforeseen circumstances. Practice living off-grid on a smaller scale before making the full transition to ensure you're comfortable with the lifestyle.

10. **Execute Your Move:** Once you've completed your preparations, it's time to execute your move to off-grid living. Gradually transition your utilities, disconnect from centralized services, and start relying on your self-sustaining systems for energy, water, and food.

By following these simple steps and careful planning, you can successfully make the transition to off-grid living and enjoy the benefits of self-sufficiency, sustainability, and connection to nature.

Overcoming Initial Challenges

The initial stages of off-grid living come with unique challenges, from adapting to a new lifestyle to addressing unforeseen obstacles. Here are some common challenges and ways to overcome them:

1. **Energy Shortages:** In the beginning, you may experience energy shortages due to limited renewable energy production or inefficient systems. To overcome this challenge, consider optimizing your energy systems, upgrading equipment, and adjusting your energy usage habits to match available resources.

2. **Water Supply Issues:** Managing water supply can be challenging, especially during dry seasons or if water sources are limited. Implement water conservation measures, such as capturing rainwater, using low-flow fixtures, and recycling greywater, to maximize water efficiency and reduce consumption.

3. **Food Production:** Growing your own food can be challenging, particularly if you're new to gardening or farming. Start small, focus on easy-to-grow crops, and gradually expand your garden as you gain experience. Consider alternative methods such as hydroponics or container gardening if space is limited.

4. **Isolation and Loneliness:** Off-grid living can be isolating, especially if you're located far from neighbors or urban areas. Stay connected with friends and family through regular communication, join local off-grid communities or online forums, and participate in community events to combat loneliness and isolation.

5. **Maintenance and Repairs:** Maintaining off-grid systems requires regular upkeep and occasional repairs. Learn basic maintenance skills, keep spare parts on hand, and schedule routine inspections to identify potential issues before they become major problems. Consider forming a maintenance schedule or sharing responsibilities with other community members to lighten the workload.

6. **Weather Extremes:** Weather extremes such as storms, wildfires, or extreme temperatures can pose significant challenges to off-grid living. Prepare for emergencies by creating a disaster readiness plan, stockpiling emergency supplies, and fortifying your home and property against extreme weather events.

7. **Financial Constraints:** Off-grid living can be expensive to set up initially, and ongoing costs for maintenance and upgrades may strain your finances. Budget carefully, prioritize essential investments, and explore cost-saving measures such as DIY projects or bartering with neighbors for goods and services.

8. **Regulatory and Legal Issues:** Depending on your location, off-grid living may be subject to regulatory restrictions or zoning laws. Research local regulations, obtain necessary permits or approvals, and consult with legal experts if you encounter any legal issues or challenges.

Financial Planning for Off-Grid Transition

Transitioning to off-grid living requires careful financial planning to ensure a seamless and sustainable shift. Here are some steps to consider:

1. **Assess Your Current Finances:** Start by evaluating your current financial situation, including income, expenses, savings, and debts. Determine how much you can allocate towards transitioning to off-grid living and ongoing expenses.

2. **Set a Budget:** Create a comprehensive budget that outlines the costs associated with transitioning to off-grid living, including purchasing land, building or renovating a home, installing renewable energy systems, water management systems, and acquiring necessary supplies and equipment. Consider both one-time expenses and ongoing costs.

3. **Research Costs:** Research the costs of off-grid living essentials such as solar panels, wind turbines, water tanks, composting toilets, and gardening supplies. Get quotes from suppliers and contractors to estimate the total cost of your off-grid setup.

4. **Identify Funding Sources:** Explore various funding sources to finance your off-grid transition, such as personal savings, loans, grants, crowdfunding, or partnerships. Consider the pros and cons of each option and choose the ones that best suit your financial situation and goals.

5. **Prioritize Investments:** Prioritize investments based on your needs, preferences, and available resources. Determine which off-grid systems and amenities are essential for your comfort, safety, and sustainability, and allocate funds accordingly.

6. **Plan for Contingencies:** Be prepared for unexpected expenses or setbacks by setting aside a contingency fund to cover emergencies, repairs, and unexpected costs. Aim to have a buffer of at least 10-20% of your total budget to handle unforeseen challenges.

7. **Seek Cost-saving Opportunities:** Look for cost-saving opportunities to reduce expenses and maximize the value of your off-grid investment. Consider DIY projects, purchasing used or refurbished equipment, negotiating with suppliers, and exploring alternative solutions to save money.

8. **Monitor and Adjust:** Regularly monitor your financial progress and adjust your budget as needed to stay on track with your off-grid transition goals. Review your expenses, income, and savings regularly to ensure that you're meeting your financial objectives and making progress towards self-sufficiency.

9. **Plan for Long-term Sustainability:** Consider the long-term financial sustainability of your off-grid lifestyle by factoring in ongoing maintenance, repairs, and replacement costs for off-grid systems. Plan for future upgrades or expansions as needed to improve efficiency and reliability.

10. **Seek Financial Advice:** If you're uncertain about navigating the financial aspects of transitioning to an off-grid lifestyle, it may be beneficial to consult with a financial advisor or consultant

experienced in sustainable living and off-grid practices. Their tailored advice can assist you in reaching your financial objectives with confidence.

Chapter 21
Future-Proofing Your Lifestyle

Embracing an off-grid lifestyle is not just a choice; it's a commitment to self-sufficiency, sustainability, and resilience. In this chapter, we delve into the strategies for future-proofing your off-grid lifestyle. From keeping up with innovations to preparing for long-term sustainability and adapting to environmental and climate changes, this chapter will empower you to thrive in the evolving landscape of off-grid living.

Keeping Up with Innovations

Embracing a sustainable and self-reliant off-grid lifestyle demands a continuous commitment to staying abreast of the latest innovations. By remaining vigilant about technological advancements, you not only enhance the efficiency of your systems but also contribute to minimizing your environmental footprint. Here's how to keep up with innovations in simple steps:

1. **Stay Connected:** Join online communities, forums, and social media groups focused on off-grid living, renewable energy, and sustainable technology. Engage with like-minded individuals, share experiences, and exchange information about the latest innovations and best practices.

2. **Research and Learn:** Take the time to research and learn about emerging technologies relevant to off-grid living, such as advances in solar panels, wind turbines, energy storage systems, water purification technologies, and sustainable agriculture practices. Stay updated on industry news, publications, and research findings to stay informed about the latest developments.

3. **Attend Events and Workshops:** Attend industry events, workshops, and conferences dedicated to renewable energy, sustainability, and off-grid living. These events provide opportunities to learn from experts, discover new technologies, and network with peers in the field.

4. **Follow Industry Leaders:** Follow influential figures, experts, and thought leaders in the off-grid and sustainability space on social media, blogs, and other platforms. Keep an eye on their insights, recommendations, and updates about emerging technologies and trends.

5. **Subscribe to Newsletters and Publications:** Subscribe to newsletters, magazines, and publications focused on renewable energy, sustainability, and off-grid living. Receive regular updates, articles, and insights about new technologies, product reviews, and case studies.

6. **Participate in Demonstrations and Trials:** Participate in technology demonstrations, pilot projects, or field trials of emerging technologies relevant to off-grid living. Collaborate with technology providers, research institutions, or government agencies to test and evaluate new products, systems, or solutions in real-world off-grid environments.

7. **Experiment and Innovate:** Be open to experimenting with new technologies and innovative solutions in your off-grid setup. Test out new products, tools, or methods to see how they can improve efficiency, sustainability, and resilience in your lifestyle.

8. **Seek Professional Advice:** Consult with experts, consultants, and advisors specializing in off-grid living and sustainable technology. Seek their advice, recommendations, and assistance in evaluating and implementing emerging technologies tailored to your specific off-grid needs and requirements.

Preparing for Long-Term Sustainability

Embracing a lifestyle of off-grid living entails more than just stepping away from the grid – it's about creating a sustainable, resilient, and self-sufficient existence that respects the environment and secures our future. To achieve this, we must lay the groundwork for long-term sustainability through thoughtful planning and action. This section delves into the key aspects of preparing for long-term sustainability in off-grid living, offering insights and actionable strategies to fortify your self-sufficient journey.

1. **Invest in Renewable Energy:** In order to generate electricity that is both clean and sustainable, it is important to make use of renewable energy sources such as solar panels, wind turbines, and hydropower. Invest in high-quality, efficient systems that can meet your long-term energy needs and reduce reliance on fossil fuels.

2. **Implement Water Management Systems:** Set up water management systems for harvesting rainwater, recycling greywater, and conserving water usage. Install water tanks, filters, and pumps to collect and distribute water for drinking, irrigation, and household needs. Prioritize water conservation to ensure a sustainable and reliable water supply for the long term.

3. **Establish Sustainable Food Production:** Focus on sustainable food production methods such as organic gardening, permaculture, or aquaponics to grow your own food. Cultivate a diverse range of crops, fruits, and vegetables to ensure nutritional diversity and resilience against pests and diseases. Incorporate composting and soil conservation practices to improve soil health and fertility for long-term sustainability.

4. **Practice Resource Conservation:** Adopt resource conservation practices to minimize waste, reduce consumption, and maximize efficiency. Implement energy-saving measures such as insulation, energy-efficient appliances, and passive solar design to reduce energy consumption and lower utility costs. Reduce water waste by fixing leaks, using water-efficient fixtures, and practicing responsible water usage habits.

5. **Diversify and Strengthen Community Connections:** Foster strong connections with your off-grid community and neighboring communities to share resources, knowledge, and support. Collaborate on community projects, share surplus produce, and participate in collective initiatives to strengthen resilience and self-sufficiency.

6. **Stay Informed and Adaptive:** Stay informed about emerging technologies, innovations, and best practices in off-grid living and sustainability. Keep up-to-date with industry news, research findings, and community developments to adapt and evolve your off-grid lifestyle accordingly.

7. **Prepare for Environmental Challenges:** Anticipate and prepare for environmental challenges such as extreme weather events, climate change, and natural disasters. Implement resilient design strategies, reinforce infrastructure, and develop emergency preparedness plans to mitigate risks and ensure continuity in off-grid living.

8. **Invest in Education and Skills:** Continuously invest in education and skills development to enhance your self-sufficiency, resilience, and adaptability in off-grid living. Learn new skills such as gardening, renewable energy installation, and emergency response techniques to empower yourself and your community for long-term sustainability.

Adapting to Environmental and Climate Changes

Navigating the challenges of off-grid living in the face of environmental and climate changes requires a proactive approach to resilience and sustainability. Here are some strategies to adapt to these changes:

1. **Monitor Environmental Conditions:** Stay informed about local environmental conditions, such as temperature fluctuations, rainfall patterns, and natural hazards. Use weather forecasts, climate data, and environmental monitoring tools to anticipate changes and plan accordingly.

2. **Implement Resilient Design:** Design and build resilient infrastructure and systems that can withstand environmental stresses and extreme weather events. Incorporate features such as reinforced buildings, flood-resistant structures, and elevated water storage to minimize damage and ensure safety.

3. **Diversify Food Production:** Diversify your food production methods and crops to adapt to changing climate conditions. Explore resilient crops, drought-resistant varieties, and adaptive agricultural practices that can thrive in your local climate and soil conditions.

4. **Enhance Water Management:** Enhance your water management systems to cope with variations in rainfall and water availability. Invest in rainwater harvesting, water storage, and efficient irrigation techniques to conserve water and ensure a reliable supply for drinking and irrigation.

5. **Strengthen Community Connections:** Strengthen connections with your off-grid community and neighboring communities to share resources, knowledge, and support during environmental challenges. Collaborate on community resilience projects, emergency response plans, and mutual aid networks to enhance collective resilience.

6. **Practice Sustainable Land Management:** Practice sustainable land management techniques such as soil conservation, reforestation, and erosion control to protect natural ecosystems and mitigate environmental degradation. Implement permaculture principles and regenerative agriculture practices to restore soil health and biodiversity.

7. **Prepare for Extreme Events:** Prepare for extreme weather events such as storms, floods, wildfires, and heatwaves by developing emergency preparedness plans and evacuation strategies. Stockpile emergency supplies, reinforce infrastructure, and establish communication protocols to ensure safety and resilience during emergencies.

8. **Adapt Energy Systems:** Adapt your renewable energy systems to cope with fluctuations in weather and energy production. Incorporate backup power sources such as battery storage, generators, or hybrid systems to maintain energy supply during periods of low renewable energy production.

9. **Invest in Climate-Resilient Technology:** Invest in climate-resilient technologies such as passive solar design, energy-efficient appliances, and off-grid communication systems that can withstand environmental stresses and minimize energy consumption.

10. **Stay Informed and Adaptive:** Stay informed about climate change impacts, adaptation strategies, and best practices for off-grid living. Continuously monitor environmental changes and adapt your lifestyle, systems, and practices accordingly to maintain resilience and sustainability in a changing climate.

Conclusion

Reflecting on the No-Grid Journey

The journey towards off-grid living, or the no-grid lifestyle, is not just a physical transition from reliance on centralized utilities to self-sufficiency. It is a profound transformation that encompasses a shift in mindset, lifestyle, and values. As we reflect on this journey, we realize that it's not merely about disconnecting from the grid but reconnecting with ourselves, our communities, and the natural world around us.

Embracing the no-grid lifestyle involves a deep appreciation for simplicity, sustainability, and resilience. It requires us to reevaluate our priorities, reconsider our consumption patterns, and rediscover the beauty of living in harmony with nature. Throughout this journey, we have learned valuable lessons about resourcefulness, adaptability, and the interconnectedness of all living beings.

One of the most significant aspects of the no-grid journey is the sense of empowerment that comes from taking control of our own lives and destinies. By harnessing renewable energy, conserving water, growing our own food, and practicing self-reliance, we have gained a newfound sense of autonomy and freedom. We have become architects of our own futures, shaping our destinies with each decision and action we take.

Moreover, the no-grid lifestyle has brought us closer to our communities, fostering bonds of cooperation, mutual support, and shared responsibility. In times of challenge or adversity, we have come together to help one another, pooling our resources, knowledge, and skills to overcome obstacles and thrive in the face of uncertainty. Our communities have become sources of strength, resilience, and inspiration, reminding us of the power of solidarity and collective action.

As we reflect on our journey towards no-grid living, we also acknowledge the importance of gratitude and humility. We are grateful for the abundance of natural resources that sustain us, humbled by the challenges we face, and mindful of our impact on the planet. We recognize that living off-grid is not without its difficulties, but we are committed to facing them with courage, resilience, and determination.

The Road Ahead

Looking ahead, the road to no-grid living is filled with both opportunities and challenges. As we continue on this journey, we are committed to ongoing learning, innovation, and adaptation. We recognize that the world is constantly changing, and we must remain flexible and open-minded in our approach to off-grid living.

One of the key priorities for the road ahead is to embrace innovation and technological advancements that can enhance our sustainability and resilience. Whether it's investing in new renewable energy technologies, implementing water-saving innovations, or adopting smart solutions for off-grid living, we are committed to staying informed about the latest developments and incorporating them into our lifestyles.

Furthermore, we are dedicated to advocating for policies and initiatives that support off-grid living and sustainable lifestyles. By raising awareness, promoting education, and engaging with policymakers, we can create an enabling environment that empowers individuals and communities to pursue alternative living arrangements that are in harmony with nature.

At the same time, we recognize the importance of preserving and protecting the natural environment that sustains us. As stewards of the land, we are committed to practicing responsible land management, conserving biodiversity, and mitigating our ecological footprint. We strive to leave a positive legacy for future generations, ensuring that they inherit a world that is healthy, vibrant, and resilient.

In conclusion, the journey towards no-grid living is a transformative and deeply rewarding experience. It challenges us to rethink our relationship with the environment, with each other, and with ourselves. As we continue on this path, we are guided by the principles of sustainability, resilience, and community, knowing that together, we can create a brighter and more sustainable future for all.

Appendices

A: Essential Tools and Equipment

Shelter and Infrastructure:

- **Solar Panels**: For generating electricity.
- **Wind Turbine**: For generating electricity in windy areas.
- **Battery Bank**: To store generated electricity.
- **Generator**: Backup power source for low-energy days or emergencies.
- **Rainwater Collection System**: Gutters, tanks, and filters for collecting and purifying rainwater.
- **Composting Toilet**: To manage human waste without plumbing.
- **Wood Stove**: For heating and cooking, especially in colder climates.
- **Insulation Materials**: To maintain comfortable temperatures inside shelters.

Food Production and Storage:

- **Greenhouse**: For extending growing seasons and growing food year-round.
- **Garden Tools**: Shovels, rakes, hoes, etc., for gardening.
- **Seeds and Seedlings**: For planting vegetables, fruits, and herbs.
- **Livestock**: Chickens, goats, or other animals for eggs, milk, and meat.
- **Canning and Preserving Equipment**: Mason jars, pressure canners, etc.
- **Root Cellar**: For storing root vegetables and other perishable foods without electricity.
- **Solar Dehydrator**: For preserving fruits, vegetables, and herbs.

Water Management:

- **Water Filtration System**: Filters, purifiers, and chemicals for drinking water.
- **Hand Pump or Well**: For accessing groundwater without relying on municipal sources.
- **Water Storage Tanks**: To store collected rainwater or pumped water.
- **Water Conservation Tools**: Low-flow faucets, showerheads, and greywater recycling systems.

Transportation and Mobility:

- **Bicycle**: For short-distance transportation and exercise.
- **Electric ATV or Off-Road Vehicle**: For accessing remote areas or transporting goods.
- **Hiking Gear**: Boots, backpacks, and navigation tools for exploring without relying on vehicles.
- **Hand Tools**: Axes, saws, and machetes for clearing paths or cutting firewood.

Miscellaneous:

- **First Aid Kit**: Essential medical supplies and equipment for emergencies.
- **Communication Devices**: Two-way radios, satellite phones, or CB radios for communication.
- **Fire Extinguishers**: For fire safety in remote areas.
- **Emergency Supplies**: Flashlights, candles, matches, and emergency blankets.
- **Solar Oven**: For cooking without electricity.
- **Waterproof and Insulated Clothing**: To stay warm and dry in all weather conditions.

B: Recommended Reading and Resources

- **"The Encyclopedia of Country Living" by Carla Emery**: This comprehensive guide covers everything from gardening and food preservation to raising animals and building structures. It's often considered the ultimate resource for homesteading and living off the grid.
- **"The Backyard Homestead" series by Carleen Madigan**: These books provide tips and techniques for growing your own food, raising animals, and preserving resources on a small-scale homestead.
- **"Off the Grid: Inside the Movement for More Space, Less Government, and True Independence in Modern America" by Nick Rosen**: This book explores the motivations and experiences of people who choose to live off the grid, offering insights into the challenges and rewards of this lifestyle.
- **"The Self-Sufficient Life and How to Live It" by John Seymour**: A classic guide to self-sufficiency, covering topics such as gardening, food preservation, animal husbandry, and more.
- **Websites and Online Communities**: Explore online resources such as Mother Earth News, Off Grid World, and Homestead.org for articles, forums, and community support on living off the grid.
- **Documentaries and YouTube Channels**: Watch documentaries like "Off the Grid: Life on the Mesa" or follow YouTube channels like Off Grid with Doug and Stacy for real-life examples and practical tips on living off the grid.
- **Permaculture Resources**: Explore permaculture principles and techniques through books like "Gaia's Garden" by Toby Hemenway or online courses offered by organizations like the Permaculture Research Institute.

C: Checklists and Planning Guides

Shelter and Infrastructure:

- [ ] Secure land suitable for off-grid living.
- [ ] Decide on the type of shelter (tiny house, cabin, yurt, etc.).
- [ ] Obtain necessary permits and approvals for construction.
- [ ] Install renewable energy systems (solar panels, wind turbines, etc.).
- [ ] Set up rainwater harvesting system or well for water supply.
- [ ] Establish a wastewater management system (composting toilet, septic tank, etc.).

Food Production:

- [ ] Plan and create a garden area for growing vegetables, fruits, and herbs.
- [ ] Consider livestock options (chickens, goats, rabbits, etc.) if applicable.
- [ ] Learn about permaculture techniques for sustainable agriculture.

- ☐ Set up composting system for organic waste.

Water Management:
- ☐ Install water filtration and purification systems for drinking water.
- ☐ Implement water conservation measures (low-flow fixtures, greywater reuse, etc.).
- ☐ Regularly inspect and maintain water collection and storage systems.

Energy Generation:
- ☐ Determine energy needs and size renewable energy systems accordingly.
- ☐ Install solar panels, wind turbines, or other renewable energy sources.
- ☐ Set up battery storage or alternative energy storage solutions.
- ☐ Develop a backup plan for power outages (generator, battery backup, etc.).

Heating and Cooling:
- ☐ Choose efficient heating options (wood stove, passive solar design, etc.).
- ☐ Implement insulation and weatherization techniques to maximize energy efficiency.
- ☐ Consider natural cooling methods (cross ventilation, shade trees, etc.).

Food Preservation and Storage:
- ☐ Equip your off-grid kitchen with food preservation tools (canning supplies, dehydrator, etc.).
- ☐ Create a root cellar or other storage facilities for storing food supplies.
- ☐ Learn food preservation techniques such as canning, fermenting, and drying.

Communication and Connectivity:
- ☐ Establish reliable communication methods (satellite phone, two-way radio, etc.).
- ☐ Consider internet connectivity options (satellite internet, mobile hotspot, etc.).

Emergency Preparedness:
- ☐ Develop an emergency preparedness plan for natural disasters and other emergencies.
- ☐ Stockpile essential supplies (food, water, medical supplies, etc.).
- ☐ Learn first aid and basic survival skills.

Waste Management:
- ☐ Implement a waste reduction strategy (composting, recycling, etc.).
- ☐ Dispose of hazardous waste properly (batteries, electronics, etc.).
- ☐ Keep the environment clean and minimize environmental impact.

Community and Support:

☐ Connect with local off-grid communities and support networks.

☐ Attend workshops, seminars, and events related to off-grid living.

☐ Share knowledge and resources with like-minded individuals.

Index

Advanced Energy System:, 153
Advanced Water System:, 153
A-Frame Shelter, 29
Allergic Reactions, 140
Basic Self-Defense Moves, 49
Berkey Water Filter, 44
Biomass Energy, 120
Boiling, 86
Bow Drill, 23
Building a Greenhouse:, 150
Building a Simple Chicken Coop, 65
Building an Aquaponics System:, 151
Burns, 140
Bushcraft Knife, 31
Camp Axe, 32
Catchment Ponds, 85
Charcoal Filtration, 88
Choking, 141
Cisterns, 83
Cloth Filtration, 87
Clothing and Gear, 47
Compass Navigation, 34
Constructing a Small-Scale Goat Shelter, 66
Cuts and Wounds, 139
Debris Hut Shelter, 28
Desert Clothing, 48
DIY Berkey Water Filter, 90
DIY Ceramic Filter, 90
DIY Compass, 35
Edible Mushrooms, 76
Edible Plants, 74
Egg Processing, 71
Fire Piston, 25
Fire Steel and Flint, 26

Foraging and Wildcrafting, 39
Fractures and Sprains, 140
Gardening and Permaculture, 38
Greywater System, 95
Hand Drill, 24
Heart Attack, 141
Heat Exhaustion and Heatstroke, 140
Hydroelectric Energy, 119
Improvised Weapons, 50
Lean-to Shelter, 27
Map Reading, 33
Meat Processing, 70
Milk Processing, 71
Mountain Shelter, 48
Natural Springs and Wells, 42
Pepper Spray or Mace, 51
Personal Alarm System, 50
Personal Defense Tool, 51
Rain Barrels, 82
Rainwater Collection and Storage, 43
Rainwater Harvesting System, 41, 97
Raised Bed Garden, 40
Sand and Gravel Filtration, 88
Seizures, 141
Setting Up a Small-Scale Rabbit Hutch, 67
Shelter-Building, 46
Solar Disinfection (SODIS), 89
Solar Still, 45
Sun Dial, 36
Tropical Shelter, 47
Utilizing Byproducts, 73
Wind Energy, 118
Wool Processing, 72

Made in United States
Troutdale, OR
09/14/2024

22792058R00075